T0214752

Studies in Computational Intelligence

Data, Semantics and Cloud Computing

Volume 759

Series editor

Amandeep S. Sidhu, Biological Mapping Research Institute, Perth, WA, Australia
e-mail: dscc@biomap.org

More information about this series at http://www.springer.com/series/11756

Rong Kun Jason Tan · John A. Leong
Amandeep S. Sidhu

Optimized Cloud Based Scheduling

 Springer

Rong Kun Jason Tan
Curtin Sarawak Research Institute
Curtin University
Miri, Sarawak
Malaysia

Amandeep S. Sidhu
Biological Mapping Research Institute
Perth, WA
Australia

John A. Leong
Curtin Sarawak Research Institute
Curtin University
Miri, Sarawak
Malaysia

ISSN 1860-949X   ISSN 1860-9503   (electronic)
Studies in Computational Intelligence
Data, Semantics and Cloud Computing
ISBN 978-3-030-10333-0   ISBN 978-3-319-73214-5   (eBook)
https://doi.org/10.1007/978-3-319-73214-5

Printed on acid-free paper

This Springer imprint is published by Springer Nature
The registered company is Springer International Publishing AG
The registered company address is: Gewerbestrasse 11, 6330 Cham, Switzerland

Contents

List of Figures

List of Tables

List of Tables

List of Algorithms

Chapter 1
Introduction

1.1 Project Overview

Recent research has noted several trends in IT. First, there has been an increased application of IT across many sectors [1]. Continued innovation drives the need for better IT flexibility as new and more intensive uses are found for computational processing [2]. A corporation's IT solution needs to constantly be ready to adapt to new utilizations.

Secondly, there has been an exponential increase in data generation [1–3]. With increasingly large amounts of data being generated and requiring processing, IT capabilities must be able to scale easily to match the ever-growing IT demands [4].

The third trend is growing device proliferation [5, 6]. As device technology improves, available bandwidth and usage of internet-connected mobile devices has increased. Everyone now has some form of electronic terminal with various forms, sizes and levels of processing power. A majority of these ubiquitous devices generate data. Moreover, as sensor technology improves and more data is digitized, even small scale applications may generate huge datasets.

In the past 20 years, the tsunami of data that has been produced by the non-stop rise of computational power has led to a paradigm shift in large scale data processing mechanisms and computing architectures [7–9]. Jim Gray, a Microsoft researcher and database software developer, calls this the "fourth paradigm" [8].

As computational systems shift towards this "fourth paradigm", research institutions need to process more data-intensive operations which require more human and infrastructural resources. Naturally, this leads to higher storage and management expenses. This is especially noticeable in scientific research where data growth is particularly vigorous.

Consequently, the IT infrastructure upgrade life cycle of an organization is constantly shortening to meet evolving computational demands. Additional human and infrastructural resources need to be assigned to meet the ever-changing data processing requirements, causing high storage and management expenses [10].

© Springer International Publishing AG 2018
R. K. J. Tan et al., *Optimized Cloud Based Scheduling*, Studies in Computational Intelligence 759, https://doi.org/10.1007/978-3-319-73214-5_1

The trends stated above and the challenges they pose to traditional IT solutions (a growing need for better adaptability, scalability and compatibility) requires an organized and standard approach to address the challenges of this new, data-rich, world with an architecture that is able to scale into the predictable future. There is now a need for a unified solution for data creation, processing, storage and access that is freely scalable and accessible by the multitude of available devices.

A possible solution is by building our IT capabilities on cloud computing provided through IT-as-a-service (ITaaS) [11–13]. ITaaS works by providing a holistic coverage [14, 15] for an enterprise's software needs through a vendor. The vendor is responsible for managing infrastructure upgrades, negating costly overhead expenditure and thus greatly reducing expenses. The vendors are capable of offering ITaaS through cloud technology which entails the virtualization of traditional datacenters, which enables them to freely scale multiple virtual infrastructures within their cloud servers [16].

The flexibility of ITaaS is a huge boon for researchers that use large-scale data computing. The computing needs of a researcher vary greatly over time, as large spikes in computing power demand only occur over the duration of a project and the peak amount needed is highly dependent on the project itself. A temporary boost in processing power required for a single project or dataset computation would probably be unavailable or infeasible in a traditional computing environment because of the inability to justify the resources needed to upgrade a pre-existing computing cluster. If the organization is big enough, multiple projects may share a computing cluster, thus justifying the cost but in turn resulting in wait times while upgrades are in progress and as other users use the cluster. By using ITaaS to perform High Performance Computing, users can overcome the aforementioned issues with ITaaS's inherent flexibility and need only pay for the computing power they use, when they use it.

Cloud computing [17, 18] is a rising edge Internet based computational model which provides on-demand and pay-per-use basis services to subscribers. It is a new managing way for a company without involving a revolutionary change in technologies to handle the workflow of IT department if compared to grid computing [19]. Many researches [20–25] have been done to involve cloud computing in High Performance Computing (HPC) due to its optimal network latency, computational ability and scalability. Hence, in terms of scalability the cloud can be scaled from one data centre to multi-data centres with different geographical administrative domains and heterogeneous resource management model to be a federated cloud system [26].

Infrastructures as a Services (IaaS) is one of the three classifications in clouds which has the most important role because the main function is to host the hardware and associated services such as VM virtualization [27] to run a cloud including both PaaS and SaaS. Due to the importance of IaaS, task scheduling especially data intensive tasks such as networking data [28], computational biology data [29], astronomy data [30], high-energy physics [31], earth science data [32], business data [33] and others all have a unified addressing called Big Data [34] which require multiple geo-distributed parties to perform data mining in collaboration.

Data intensive tasks are highly possible to be dynamically distributed widely with adapting variety of user requirements, limitation of physical infrastructure and real time dynamic modification which all are difficult to predict in advance. Efficient task scheduling algorithm is necessary to be developed and implemented to prevent profit loss, performance decline, high processing time, low utilization and even service downtime. The platform managed by IaaS provider should focus on scheduling strategies [35] when the resources are not in satisfying level which are determined by optimization goals such as whether to maximize profit gain or service performance for increasing reputation from customers. Besides, an optimal scheduling algorithm should produce the optimal results to protect organizational benefits which is not as complex as NP-hard problems [36].

There is a rapid development of multi-core processors which the major chip vendors has a massive and constant breakthrough in increasing number of cores per processor [37]. As the result, tasks in sequential processing has no comparable performance to compete as novel as parallel batch processing, leading to shorter execution time and unleash the full potential of multi-core processor utilization.

MapReduce is an example of cluster-based single key batch processing model [38] which divide data into N number of small chunks, master node assigns idle sub nodes into two groups, one will be doing mapping single key tasks and another group doing reduce single key tasks during next phase. Once the mapping group finish the tasks, the reducing group will remotely access the data from data storage of each node of mapping group to perform reduce tasks and output N number of results as the input of next round.

1.2 Project Description

As the increasing number of service users, the IaaS service provider faces difficulties in proper distributed of limited available resources. In this situation, they will be requested to process variety type of tasks as less consumed time as possible. This is a tough challenges to the service providers to adapt random environment. Service users nowadays expect an OS-independent service because different OS will be implemented in their business as long as it fits their business model and benefit gains. Thus, platform independent design was taken into consideration.

For optimizing online scheduling in IaaS, researches on comprehensive fields have been proposed and the most notable factors are to increase the profit gains and reduce the time span while maintaining the service performance based on service level agreements (SLA) between IaaS provider and service users. These are definitely the limitation especially defining the balance point to achieve scalability in heterogeneous multi-cloud platform task scheduling with automated running process.

In this project, a scalable scheduling algorithm was being designed to determine the efficiency and effectiveness of proposed optimization of online scheduling algorithm was an optimal solution.

The concept of algorithm is tree-based parallel batch scheduling design with platform-independent scalable levels from multi-cores processing to various heterogeneous IaaS clouds processing to accomplish the data intensive tasks given by service users.

1.3 Project Objectives

An effective algorithm is core operation for IaaS task scheduling to provide on-demand service and automated scheduling algorithm is able to boost the quality of service within the scope of service level agreement (SLA) and process the task smoothly without any serious error. All of it result the key success of IaaS service providers.

Therefore, in this project, the main objectives include:

- Development of dynamic data partitioning algorithm.
- Development of four-phase core-based multi-key parallel batch processing model which all designated cores are fully utilized with load-balancing mechanism during processing period of each phase.
- Development of local memory sharing process in four phase model without accessing external data storage in iteration.
- Development of tree-based scalable processing model with most optimal output result to service users.
- Benchmarking on current prevailing public clouds.
- Study the backgrounds of scalable data-agnostic processing model with priori scheduling and implementation of big data computation for the cloud.

1.4 Motivations and Significances of the Project

The motivation of this project is the significant combination of processing big data tasks in cloud and the details are explained as follows. Cloud [17] is a cluster of computing resources including both hardware and software over a highly available network infrastructure and the origin of the cloud symbol was to represent internet since 1994. Cloud computing is to utilize the resources of cloud and provide various on-demand services over a network instantly. It is a new managing way for a company without involving a revolutionary change in technologies to handle the workflow of IT department if compared to those traditional ways. One of the example is Grid Computing which was referred to the computing resources in separated location as all are integrated in the distributed network. Cloud computing enhanced the concept because it overcomes the major drawback of Grid Computing which is single point of failure make the whole grid performance degrade. Without single point of failure, it is able to realize the achievement to magnify cost savings,

Table 1.1 Comparison between grid and cloud computing

Category	Grid computing	Cloud computing
Operating concept (computing power and data storage)	Shared	Leased
Service payment	Government or public funded	User pays to cloud provider and cloud provider pays to cloud resource provider
Location of computing resources	Distributed across different sites, countries and continents	Cloud provider's data centres are centralized in few locations with excellent network connection
Duration of jobs and access speed	Limited, it depends on condition due to different geographical location	Long terms, instant and on-demand services
Major drawback	High complexity	Data confidentiality

Table 1.2 Comparisons of IAAS, PAAS and SAAS

Category	Infrastructures as a Service (IAAS)	Platforms as a Service (PAAS)	Software as a Service (SAAS)
Functions	Hosts and provides hardware, software and services for operating cloud environment	Hosts and provides computing platform and associated solution stack access	Hosts and provides software applications access
Supplier	Cloud resources provider	Cloud services provider	Software services provider
Target user	Cloud services provider	Software services provider	Software user
Service delivery	Internet	Internet	Internet
Advantages	Online, high availability	Online, high availability	Online, high availability

availability, scalability and flexibility. The comparisons between Grid computing and Cloud computing as shown in Table 1.1.

Cloud Computing was generally classified in Table 1.2.

A Five Tier Cloud Computing Structure is shown on Fig. 1.1 and the details are shown on Fig. 1.2

1. Cloud Resources Provider offers IAAS.
2. Cloud Computing Environment contains of the cloud services offered by Cloud Services Provider and Software Services Provider.
3. Cloud Services Provider has a dependency on Cloud Resources Provider as it utilizes IAAS and provides PAAS.
4. Software Services Provider has a dependency on Cloud Services Provider as it utilizes PAAS and provides SAAS.

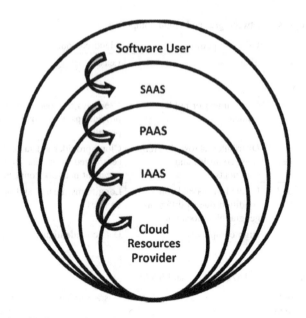

Fig. 1.1 Five tier cloud computing structure

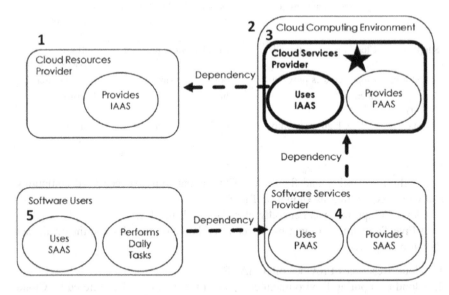

Fig. 1.2 Details of cloud computing structure

5. Software User has a dependency on Software Services Provider as it utilizes SAAS and performs daily tasks.

Note: The main important challenge is how the Cloud Services Providers utilize the IAAS and to provide PAAS as optimal services as shown on Fig. 1.2.

Big data [28] is a changing term which is diversely defined due to its popularity and even difficult to make a consensus to standard definition. It is because the definition of big data is not only based on the concept of massive volume of data but it is also categorized into three sections as follows:

- **Attributive Definition**: Big data technologies is an emerging technologies to be cost-effectively extract precious value from massive amount of a wide variety of data with high speed capturing, effective discovery and accurate analysis. This was described by IDC [39], a pioneer in big data research, in a year 2011 report to explain the "4Vs" characteristic of big data which are volume, variety, velocity and value. Volume is actual and complete size of the datasets. For instance, big data is defined from Megabyte to Gigabyte in 1970s, Gigabyte to Terabyte in 1980s, Terabyte to Petabyte in 1990s and Petabyte to Exabyte under current development trends. Variety is the form of data. For instance, structured data, unstructured data and semi-structured data. Structured data such as relational database to keep the relationship information of one piece of data and the other. Unstructured data such as documents, videos, sounds or images with no standard format and sequence. Semi-structured data is between the aforementioned both such as XML lacks of formal structure but with tags for easier organization. Velocity is the rate to match the speed of data processing. For instance, time-sensitive operation such as fraud detection must be processed as fast as possible. Value is the commercial benefits earned by using effective Big Data mining techniques.
- **Comparative Definition**: Big Data is beyond the ability of traditional data management software to capture in details, store completely, manage under control and analyse overall. This concept was brought by Mckinsey's report [28] in year 2011. A comparison Table 1.3 between traditional data and Big Data as follows:

Table 1.3 Comparisons between traditional and big data

Categories	Traditional data	Big data
Data volume	Gigabytes	Petabytes and above
Data generated rate	Hourly and above	Fast interval
Data source	Centralized	Distributed
Data integration	Easy	Difficult
Data structure	Structured	Semi-structured or unstructured
Data access time	Processing time needed	On-demand

- **Architectural Definition**: Big Data limits the effective analysis if traditional relational approaches are being used. On the other hand, significant horizontal scaling might be required to process it efficiently. This [40] is suggested by The National Institute of Standards and Technology (NIST). Besides, Big Data can be further divided into Big Data Framework and Big Data Science. Big Data Framework contains the software libraries with algorithms for Big Data processing operations. Big Data Science contains the study of capturing, conditioning and analysis techniques of Big Data. Big Data Infrastructure is the instance of Big Data Framework.

When it comes to exponential data growth in scientific circles, two fields in particular have stood out in generating ever-growing datasets that require processing. These fields are genomics, particularly gene sequencing, and astronomy.

As an example from the field of genomics, the rise of Next Generation Sequencing (NGS) in the last decade alone has seen an increase in the amount of data generated from roughly 10 Mbases per day to 40 Gbases per day on a single sequencer, i.e. a 4000% increase in generated data per sequencer; and there are multiple sequencing labs running multiple sequencers world-wide [41] driving up the increase of generated data even further.

The evolution in genomic data generation is only matched by the astronomical increase in data generated by the field of astronomy. An isolated example would be the Low Frequency Array (LOFAR), which is a European radio telescope which can create up to 5 PB of raw data over an average observation time of only 4 h [42]. When the Large Synoptic Survey Telescope (LSST) in northern Chile comes online on 2022, it is predicted to generate 30 TB per night [43]. Another astronomy project coming online is the Square Kilometer Array (SKA) to be constructed in Africa and Australia that is supposed to come online in 2020 and will generate about 2740 TB a day [44].

Table 1.4 Outline

Chapter 1	This chapter is mainly project introduction which includes project overview, description, objectives, motivations and significances
Chapter 2	This chapter focuses on project background and literature reviews related to project
Chapter 3	This chapter discusses about benchmarking on current prevailing public clouds
Chapter 4	This chapter introduces the backgrounds and work done to enable computation of large datasets
Chapter 5	This chapter introduces the details of proposed optimized online scheduling algorithm
Chapter 6	This chapter discusses about the complexity analysis and experimental results of project
Chapter 7	This chapter is the conclusion of the project about giving an idea of overall review and discussion on the future works to be researched

By digitizing all that raw data for storage, it is clear that the storage required would need to grow at the same rate as the volume of data to accommodate advances in the field. The increase in data generation also means that there is a lot of data that needs to be computed which in turn may generate more data for further processing. The exponential increase in required computing power can only be supported if advances in computing power progress at an equivalent rate or new computing techniques are developed.

1.5   Outline

Seven main chapters will be discussed in Table 1.4.

Chapter 2
Background

2.1 Literature Reviews

2.1.1 Scale of Things—Peta and Exascale

When measuring large data volumes and computations, working in the region of petabytes and petaflops is considered working in petascale (10^{15}). Supercomputers achieved the feat of being able to compute at a petaflop in 2007, as listed in the Top500 List which ranks and keeps track of the 500 fastest computers annually [45]. The next evolutionary step would be to go up another order of magnitude to the exascale (10^{18}).

This increase in scale is inevitable as data grows but the jump from petascale to exascale faces many challenges. The primary issue is that there is precious little time to make the move to exascale. The rate of increase in data generation is growing much faster than advances in computing power and data storage capabilities. This is most easily summarized in Fig. 2.1, which is a graph by Kahn [41] charting the growth of genomic data (generated from genetic sequencing), compared to growth in computing and storage capabilities of computing infrastructure.

The rate of growth in raw data generation outstrips both Moore's Law and Kryder's Law which estimates the rate of growth in computer processor power and storage capabilities respectively. This means that raw data is generated faster than advancements in computing technology needed to process it, a phenomenon recognized by both Kahn and Stein in their research [41, 46]. This is especially true after 2004 when Next Generation Sequencing was introduced, greatly boosting data output as the technology was gradually adopted and became more widespread. It is highly probably that other scientific fields may also develop technologies that would significantly increase raw data generation. Following these trends, the scientific community is going to require exascale computing capability before our current technology is prepared to provide it.

© Springer International Publishing AG 2018

R. K. J. Tan et al., *Optimized Cloud Based Scheduling*, Studies in Computational Intelligence 759, https://doi.org/10.1007/978-3-319-73214-5_2

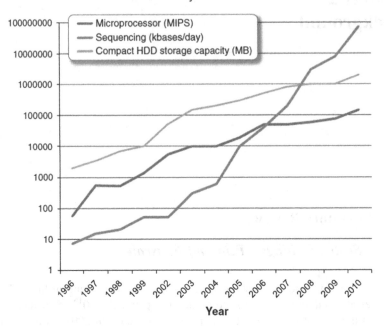

Fig. 2.1 Sequencing progress versus compute and Storage

2.1.2 Current Resource Availability

Besides increasing a processor's processing power, computer researchers have also been increasing the number of processors and storage disks on a single machine to mitigate the ever widening gap between computing requirements and the capabilities of individual computing components.

This has resulted in the multicore and server based architecture at the heart of many supercomputers, High Performance Computing clusters, and networked computer clusters used in the majority of today's scientific research. This traditional approach to scaling up computing performance is well established and a steady progression of technological advances in this traditional computing method to further its efficiency can be observed in the works of Banalagay et al. who applies a resource estimation system to reduce resource wastage when utilizing traditional supercomputers [47] and da Silva et al., who fine-tunes data processing strategies for large datasets [48] that are processed by a traditional supercomputer.

As the march towards exascale plods inexorably onwards, we take stock of the currently available systems and infrastructures we have in place.

2.1.3 High Performance Computing

High performance computing (HPC) system is a networked clustered computing system and their derivatives are the most widely accepted solution to big data computing. A HPC system can be a single supercomputer at a specialized facility, but normally consists of clustered computer systems networked together to enable them to perform computationally intensive jobs as a whole. In its most basic form, a HPC cluster would consist of a head or controller node, which manages the tasks and scheduling for the entire system, and multiple compute nodes, that would perform the actual processing. The most common of these implemented systems would be a local HPC cluster and the networked, grid, HPC cluster [49].

A system middleware allows the HPC system to capture the cumulative processing power off all nodes on the cluster [50]. Clusters are created to take large programs and sets of data and break them down into smaller computing jobs, these jobs are sent to individual compute nodes for processing. The cluster, though loosely coupled appears to work as a single computer solving processing jobs that computationally intensive [51]. Clusters are deployed to solve computationally intensive processing jobs like simulation, data analytics, web services, data mining, bioinformatics and many others [52]. For such jobs clusters improve speed and/or reliability over that provided by a single super computer, while typically being much more cost-effective than using a single super computer.

In a HPC cluster system only the head node can be accessed, which acts a scheduler [53]. This head node acts as a gateway between the user and the cluster. The head node is setup as a launching point for jobs running in the cluster [54]. Actual processing is done on the compute nodes that are connected to the head node.

Depending on the type of job sent to the cluster, processing is done using either in pipeline flows or sweep flows [55]. In a HPC cluster processing jobs that execute in a sweep flow are jobs that are divided in into threads that can be totally executed in parallel without any communication between the cluster nodes that execute each thread. Sweep flow jobs are also known as embarrassingly parallel jobs.

Other processing jobs in cluster may execute in a pipeline flow. Pipeline flows are processing jobs that have threads that have dependencies with each other and typically must be executed sequentially. However there exceptions to that, in that a particular job may partially pipelined in nature. This is when a stage in the pipeline (a processing node) completes partial results of a set of data. These partial results will be then sent to the next processing node which will begin processing of the partial results immediately. While the previous node will continue to process the remaining set of data in parallel.

Job threads are distributed among nodes and pipeline flow jobs need to communicate between nodes and between processes, both pipeline flow nodes and sweep flow nodes in a HPC cluster will also have to communicate with the head node to submit output of the processing. Due to the all the communication needed, latency becomes an issue in a HPC cluster. At the software level parallel processing algorithms [56] and parallel processing middleware frameworks [57] have been created to optimize and reduce the amount communication overhead between nodes. However most of these software optimizations are still limited by the physical network infrastructure [58].

Currently most of the existing and legacy HPC clusters are built using Fibre Channel which uses the combination of copper wiring and fibre optic cabling, which has a high data rate [59]. The current next generation HPC clusters typically utilize a technology called InfiniBand which is a type of communications link for data flow between processors and I/O devices that offers throughput of up to 54 gigabits per second. InfiniBand is also easily scalable and supports quality of service (QoS) and failover [57]. InfiniBand has a vast advantage over Fibre Channel.

As the Fibre Channel is just a medium of transmission between network interfaces of the HPC node. Meanwhile InfiniBand bypasses the typical network interface and allows direct communications between HPC cluster nodes at CPU bus level to interconnect nodes in a HPC cluster. The main barrier to usage of InfiniBand is that it is expensive to implement. Existing HPC cluster's Layer 2 physical cabling as it needs to be dismantled [60]. Hence, typically InfiniBand is only used in high end HPC clusters or cloud data centers. One way to inexpensively utilize the low latency interconnects of InfiniBand is to utilize a cloud computing service to access more processing nodes and scale up the HPC cluster. As currently most Cloud Data Centers are built by using InfiniBand. Therefore processing HPC jobs migrated to the cloud would allow most HPC jobs to get benefit from the InfiniBand found in cloud data center networks.

2.1.4 Cloud Computing

As many researchers are discovering, the traditional method of computing data on a local computer cluster, supercomputer or our current computer networks will no longer be sufficient at the exascale level. The current architectures in our networked computer clusters at the exascale level would require unsustainable power and storage resources as noted by Ahren [61]. Viability of calculation at the exascale is much reduced as there will be compounded delays, such as data transfer times, in our current networked systems that will become non-negligible when scaled up to that level.

The IT industry has already been steadily adopting cloud computing technologies due to its cost savings and scalability [12]. Considering this, many researchers in the scientific community have been advocating a switch to cloud computing for big datasets alongside other data saving techniques. This is evident from the myriad works [9, 61–80] which notes the widening discrepancy between processing power and data and suggests implementations that apply cloud computing to solve a wide range of high performance computational tasks.

Cloud computing is slowly being established as a new paradigm for the computing infrastructure establishment. This paradigm moves the placement of the network infrastructure to lessen the expenses related to the hardware and software resources management [81]. National Institute of Standards and Technology defines Cloud computing as an architecture for enabling ubiquitous, convenient, on-demand network access to a shared pool of configurable computing resources (e.g., networks, servers, storage, applications, and services) that can be rapidly provisioned and released with minimal management effort or service provider interaction.

Cloud computing defines the use of delivering hosted services over the internet [16]. It is also a basic change to an operational architecture in which applications are not necessarily kept in a specific physical hardware. Computing resources can be easily relocated based on the user's need due to cloud's flexibility [82]. Cloud computing provides three types of service models namely in the form of Infrastructure as a service (IaaS) [83] and Platform as a service (PaaS) [84], Software as a service (SaaS) [85]. The three service models are contrasted to the traditional IT architecture which requires the organization to self-manage all IT resources [85] as previously shown on Figs. 1.1, 1.2 and Table 1.2. In cloud computing cost savings are achieved when selected IT components are managed by cloud service providers [86].

The extensive work that has been done in academia fleshes out the three cloud implementation models, namely: public [86], private and hybrid clouds. The private cloud is a more personalized set of services offered to a particular organization catering directly to the organization's needs and its resources are normally managed directly by the organization. A hybrid cloud is a combination of a private cloud that integrates additional resources from the public cloud. The public cloud model is where a vendor offers a fixed variation of cloud services to multiple users. Services provided by public cloud are either free (NECTAR) [87] or users may need to pay every time they are using the service (Azure, Amazon EC2) [88, 89]. Users need to ensure that they have disaster recovery and data backup plans. Usually the cloud is managed and run at a data center owned by the cloud service provider and is multi-tenant. This type of shared architecture reduces the costs but since the service provider owns the underlying infrastructure, the visibility and control is less in a public cloud [90].

Big ICT companies like Microsoft, Amazon and Google provide public cloud services and already have many global data centers to host cloud computing applications. The main purpose of these data centers is to maximize computing resources through virtualization and shared usage between multiple end users. The public cloud offers many advantages to users that will also benefit the large scale usage in scientific computing.

2.1.4.1   Cost Efficiency

The public cloud supplies basic elements like network bandwidth, storage and CPUs as services at cheap unit prices. Costs can be driven down thanks to the public cloud's "pay-as-you-use" model. Cloud users also need no longer be concerned about scalability since storage and computational power is virtually limitless. Overall, usage of the cloud leads to much lower initial overheads. Work done by Afgan et al. to show the numerous savings of using cloud for genomic computations showed that a genomic problem set of 45 GB took 9 h to upload at 1.5 MB/s and the cost was $5 to transfer; besides costing $20 to process in 9 h or $50 to process in 6 h respectively. Hence, a complete analysis could be performed for $25 in 15 h which lead to huge savings for smaller genome research institutions which would not normally be able to justify the cost for purchasing the hardware required [9].

2.1.4.2   Low Latency

The cloud shares a weakness with the networked grid computing as in that it takes a relatively long time to upload data to a distant server. This is counteracted by the continued growth of cloud coverage. There are an increasing number of cloud data servers around the world which reduces data transfer latency as the cloud will automatically assign tasks sent to the cloud to the geographically closest server unless specifically requested otherwise. When compared to traditional grid computing, which also suffers the same data transfer latency issue, the cloud offers a much better solution as it expands to new locations much faster than the grid ever could, as noted by Marinescu [91].

Furthermore, with all data uploaded to a single location due to more scalability, using the cloud would overcome the problem with grid computing that is forced to use multiple disparate clusters at different places when running tasks larger than a single member cluster can handle. Data transfer speeds between the cloud's virtualized servers are also much faster than between their physical equivalents found in traditional computing clusters. Being able to scale computing resources "locally" at cloud data centers overcomes many of the latency problems we currently suffer in grid computing.

2.1.4.3   Versatility and Performance

The cloud is extremely versatile due to virtualization. Any cluster configuration or network architecture can be recreated. The same computing resources can easily be reconfigured and used for a different job. The biggest advantage of the cloud is the ability to expand your computational capability almost instantaneously. The public cloud and virtualization allows users to expand their computing clusters without the need for purchasing new hardware. To increase or reduce computing capacity, users need only access a centralized control panel for the public cloud and requisition or release resources. This flexibility provides computing power when it is needed, leading to financial savings as expensive, new hardware need not be purchased to meet a temporary peak load. Besides that, requisitioning resources and setting up both virtual machines and clusters on the cloud is much faster than ordering new physical compute nodes and adding them to a HPC cluster [92]. For more specialized applications, software and data management systems can be shared among multiple research institutions to achieve higher efficiency [93].

The advantages in flexibility offered by the cloud has to be complemented by actual computing performance. Using industry standard benchmarks, a comparison of the performance of various public clouds [94] commonly used by academic organizations in the Asia Pacific Region—Microsoft Azure (http://azure.microsoft.com/), Amazon EC2 (http://aws.amazon.com/ec2/) and the Australian National eResearch Collaboration Tools and Resources Cloud (NeCTAR) (https://www.nectar.org.au/research-cloud). The resources requisitioned on all of the clusters were the same. Clean installations of virtual machines running Windows Server 2012 R2 OS were created on the cloud and the virtual machines were then virtually networked into a cluster. Linpack Benchmark software (http://www.netlib.org/benchmark/hpl/) was then installed on the head node virtual machine and the benchmarks run. The public clouds all performed similarly for most of the compute benchmarks with Windows Azure performing the best with a cluster efficiency performance of 88.82% (Table 2.1).

Theoretical maximum performance of the cluster on the cloud was calculated by summing up the rated performance of the cores requisitioned in the cloud for the benchmark. The actual cluster performance was calculated based on the speed the benchmark was performed. The cluster efficiency could then be calculated by dividing the actual cluster performance by the theoretical cluster performance.

The Public Cloud outperforms the physical based HPC clusters in terms of efficiency because the entire cluster is virtualized on the same data center. Virtualized servers also run on standardized hardware that eliminates incompatibility that may reduce performance. This makes the public cloud a very attractive alternative to current on premise or networked computing solutions. Further works that support the effectiveness of the cloud as a computing alternative to local clusters is noted by both Chaisiri and Sidhu [52, 95].

Additionally, there is no longer any hassle over creating backups because it is the cloud service provider's responsibility to replace any faulty components and create temporary copies of the data. These benefits of the public cloud enable organizations to concentrate on innovation instead of worrying about building huge data centers to meet their computational requirements.

Table 2.1 Windows Azure benchmark results		
Theoretical maximum performance	332.8 GFlops
Actual cluster performance	295.6 GFlops
Cluster performance percentage	88.82%

2.1.4.4 Advances in Cloud Services

Current cloud solutions provide Infrastructure-as-a-Service (IaaS), Platform-as-a-Service (PaaS), and Software-as-a-Service (SaaS) to users. IaaS is a provision model in which an organization outsources hardware resources such as storage, processor cores, and networking components from a cloud service provider. The client only needs to pay for the service on a per-use basis as the housing, running and maintenance of the equipment is the responsibility of the service provider. PaaS goes a step further as users rent the operating systems which is typically run on rented IaaS. SaaS is the software distribution model for additional applications to be installed on rented PaaS that is made available through a vendor. This provides the basic framework for the virtualization of IT infrastructure for nearly all commercial organizations which typically consists of hardware, operating systems and applications.

Further value can be added for users with more specialized needs by enabling custom, virtualized networks through Network-as-a-Service (NaaS). NaaS enables virtual machines running PaaS to be networked together creating the complex clusters used for scientific calculations. Another useful feature would be Desktop-as-a-Service (DaaS) which increases accessibility by enabling users to access their virtual machine operating systems from any terminal.

Researchers would greatly benefit from more research oriented services such as Data-Analytics-as-a-Service (DAaaS) and High-Performance-Computing-as-a-Service (HPCaaS), which are research specific software and hardware to increase specialized functionality on the cloud. Several public cloud providers offer specialized nodes for HPCaaS that are fine-tuned to compute-intensive, memory-intensive or network-intensive tasks depending on the needs of a project (Fig. 2.2).

Private cloud is an exclusive computing model which provides services to limited number of people behind the firewall. Some organizations (like banks) are highly concerned about the data security so they choose private clouds over public clouds [96]. Private clouds are among the least implemented of all cloud models as hardware and implementation costs can be substantial. This cloud is not really that cost-effective but it has the highest security level compared to other cloud models [97].

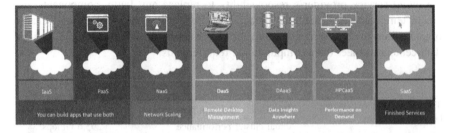

Fig. 2.2 Current cloud services

A hybrid cloud consists of one or more private cloud and one or more public cloud. It is an environment where an organization administers some resources in-house on the private cloud and the rest of resources being managed externally on the public cloud [85]. Hybrid cloud allows the organizations to increase the usage of their IT infrastructure and thus lessen their IT expenses since the local infrastructure integrates with the computing capacity from a public cloud. When the public cloud is used, the duration of load peaks is quite short and it compensates the high premium that is charged by the provider of the public cloud making hybrid cloud more cost effective in comparison to using the private cloud alone [94]. A hybrid cloud has to be built specially to overcome the latency of communications over the internet, especially in synchronizing data and processes between the private cloud and public cloud.

Hybrid cloud technology, specifically the application layer, transport layer, session layer stack optimizations that enables private cloud and public cloud to seamlessly synchronize with each other can now be applied to integrate a local on premise HPC cluster with HPC processing nodes situated on the cloud, hence the term is created, that is HPC + Cloud.

The major application of the HPC + Cloud architecture is to be able scale the HPC cluster by provisioning new virtual processing nodes from the cloud on an on demand basis. All the advantages of the hybrid cloud are inherited by the proposed HPC + Cloud paradigm but with lesser administrative overhead of a hybrid cloud since there is no need to manage a private cloud on premise. Hybrid Clouds despite their advantage in integrating the public and private cloud incur high investment in network infrastructure to assemble a private cloud on premise.

However with the HPC + Cloud architecture, there is no need to build a private cloud to enable HPC + Cloud. All required is to configure the existing HPC to provision processing nodes using the HPC + Cloud software framework that it will be described further in the next sections.

2.1.5   Existing HPC Implementation Architecture

HPC systems are clustered computer systems that are designed according to specific architectures to enable it to perform computationally intensive jobs. This section addresses the various implementation architectures that are being employed in HPC clusters.

2.1.5.1   On Premise Local HPC Cluster

The simplest HPC implementation has always been to host all the HPC nodes in a single premise [49]. And to ensure that the HPC cluster is powerful it's built using expensive compute nodes that have high number or CPU cores and plenty of computer memory. Also to ensure high speed, low latency communications

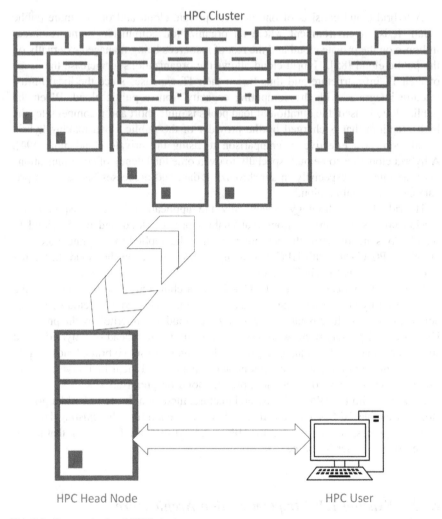

Fig. 2.3 On premise local HPC cluster

between HPC compute nodes and the head node, the nodes communicate over fiber optic networks [95]. In terms of scaling and upgrading an on premise only HPC cluster, organizations normally just purchase more nodes, upgrade CPU's, upgrade RAM or hard drive storage for each node (Fig. 2.3).

However, there are downsides to physically upgrading hardware on a HPC cluster [98]. However buying more hardware is quite ineffective as it takes time to procure new hardware and upgrade the existing HPC systems. And the demand for computing power may be immediate [88].

New hardware could also be potentially underutilized. As Most High Performance Computing clusters are normally built with peak demand in mind.

Meaning, they always try to anticipate demand spikes and make sure that the HPC system can handle that particular maximum or peak load. However demand spikes are normally only seasonal and the extra CPU and Memory resources are left underutilized in off peak seasons [99].

2.1.5.2 HPC Cluster Implemented on Grids

Grid computing is a form of distributed computing tool to deliver computing power over the internet. It consists of a grid of clusters distributed globally in various geographical locations. However using a grid computing solution require rewriting of the HPC software [100] and creating a middleware software to make it compatible with the grid computing architecture. This software rewriting and middleware creation is not a trivial exercise.

As HPC clusters on the grid uses remote clusters which are distributed all across the world. This makes latency a problem, although there are Grid software middleware available to help alleviate the latency problem, the vast geographic variability of the grid cluster is still quite a challenge to overcome.

Grid computing has the same fundamental problem for scaling hardware and software as HPC clusters. Grid doesn't solve the problem of scaling hardware and software for HPC cluster, it still requires substantial hardware and software cost to scale and upgrade. Compared to the HPC + Cloud architecture that allows hardware of the HPC to be upgraded on a demand basis thereby lowering the cost of creating a HPC cluster. Also hardware in cloud is managed by cloud service providers with a high fault tolerance and with high scalability. Moreover, a typical grid computing platforms provide no such guarantees for scalability and fault tolerance is dependent on the clusters involved (Fig. 2.4).

2.1.5.3 HPC Cluster Implemented on the Public Cloud

To take advantage of the performance of the cloud, HPC clusters consisting of both head node and compute nodes have been implemented entirely on the cloud. When implemented on a Public Cloud, scaling and upgrading the capabilities of an on premise HPC cluster is taken care of by the Public Cloud infrastructure which is very elastic in nature [101].

The main weakness of this implementation architecture is the cost of utilizing the Public Cloud alone to replace the entire existing on premise HPC infrastructure. The cost per unit of computing on cloud would be quite substantial as all current on premise local hardware and software need to be virtually replicated on the Public Cloud IaaS. Since the public cloud runs on a "pay-as-you-use" model, the resources will have to permanently be requisitioned to keep data on it for the long term. Notwithstanding the cost of discarding existing hardware of the local on premise HPC cluster.

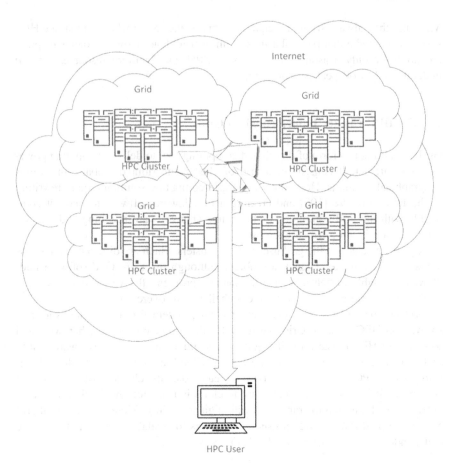

Fig. 2.4 HPC cluster implemented on grids

Due to the fact that the entire cluster processing will take place outside of the local on premise boundary. HPC applications are used to process sensitive data may face legal, regulatory, privacy or other restrictions that might make it impossible to store or process that data in the cloud. One example of this is a simulation done to predict diseases using patient medical data would be legally not allowed to leave the hospital for processing in a public cloud [102].

A HPC cluster implemented entirely on the cloud would also require high bandwidth connections between the data source which is on premise and the cloud data centre. HPC that applications typically need lots of computing power often rely on large amounts of data. So the act migrating all local on premise data to the cloud data center would require a substantial bandwidth investment [102] which will add to the cost of implementing the HPC in the cloud (Fig. 2.5).

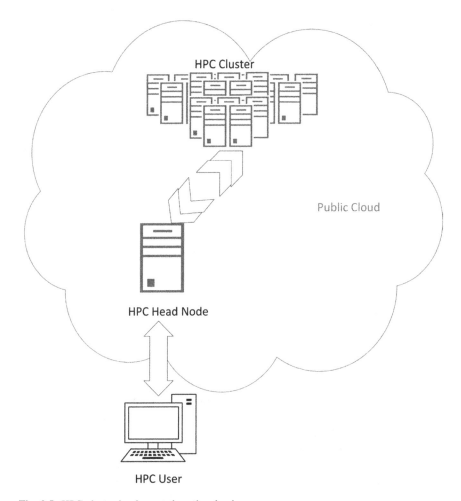

Fig. 2.5 HPC cluster implemented on the cloud

2.1.5.4 HPC Cluster Implemented Hybrid Cloud

The term "Hybrid Computing" was introduced by the journal "Hybrid Computing—
Where HPC meets grid and Cloud Computing" [86] which proposed a combination
of traditional HPCs, Grid Computing Networks and Cloud computing. This concept
consolidated the advantages, and eliminated the existing weaknesses, of all three
systems while providing a far more effective HPC workload execution method.

A hybrid cloud consists of one or more private clouds interconnected with one or
more public clouds, creating an environment where an organization administers
some resources internally and manages the rest of the resources externally on the
public cloud [86]. HPC clusters can be implemented in the Hybrid Cloud [103].

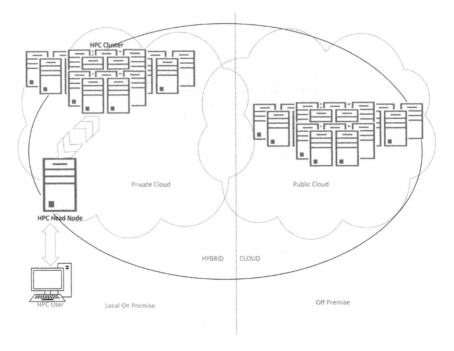

Fig. 2.6 HPC cluster implemented on the hybrid cloud

Hybrid cloud is a good solution as work load from the on premise private cloud can be seamlessly transferred to public cloud on an on demand basis and software licenses are only bought once for the private cloud (Fig. 2.6).

The main problem is the cost of creating a private cloud locally. The hardware and software investment needed to create to build a private cloud in organizations premise is quite prohibitive [104]. There is also administrative overhead of managing a private cloud as well and HPC cluster and virtual HPC nodes on the public cloud. Current hardware and software infrastructure in the existing on premise HPC cluster would have to be reconfigured to accommodate the private cloud software and hardware infrastructure.

2.1.6 Significance of HPC Clusters on Hybrid Cloud

Although a lot of work is now being done in scientific computing on managing various aspects of big data [105], research focus is still primarily on using either the public cloud or private cloud such as work done by AbdelBaky et al. to create a public cloud on a supercomputer [106] and also Afgan et al. [62] and Angiuoli et al. [63, 64].

Applying only the traditional computing method of just using local clusters brings many issues. First, local HPC clusters are inefficient since organizations cannot scale the hardware used in the cluster fast enough to meet spikes in processing demand. This forces HPC clusters to be built with overcapacity to potentially meet future spikes in resource demand. A fully private cloud solution would also share the same primary disadvantage as a traditional HPC cluster: a substantial local hardware investment is required to meet peak loading.

Hybrid clouds allows organizations to reduce the usage of local IT infrastructure and thus lessen their overhead expenses. The initial configuration of a hybrid cloud need only cater to average processing demand. During peak loads, the hybrid can be easily scaled onto the public cloud to meet demand. The duration of peak loads when the public cloud needs to be utilized is relatively short, and that compensates for the high premium on "pay-as-you-use" terms that is charged by the public cloud provider. This makes the hybrid cloud more cost effective in comparison with using the private cloud or public cloud alone.

Cloud management interfaces are centralized and more user-friendly compared to the management interfaces of local HPC clusters. The cloud's ability to scale resources and computing capability in any configuration, thanks to virtualization, allows the hybrid to efficiently distribute its workload across the cloud to be processed at multiple scalable clusters. Furthermore, the process of adding nodes into the hybrid cloud is much simpler and faster when compared to expanding a HPC cluster.

One of the main reasons for the lack of work on hybrid clouds in academic research communities can be the cost of setting up and maintaining the on premise private cloud component. This cost can be significantly mitigated by building the private cloud from a pre-existing HPC cluster in the hybrid model. Most research institutions, such as universities, already own HPC clusters and a simple upgrade of its infrastructure will provide a small private cloud that can be used in the hybrid model.

Secondly, the lack of support for the heterogeneity of the hybrid cloud and the absence of good task management or scheduling software makes using the hybrid cloud a difficult proposition at best. The complexity of creating a hybrid cloud environment means that only organizations large enough to have a dedicated IT department or support would attempt to implement it. This can rectified by providing an automated solution to setting up a hybrid cloud environment that would support the heterogeneity of the component members of the hybrid cloud while simplifying the initial implementation of the environment.

2.1.7 Hybrid Cloud Future Improvements

Adopting a hybrid cloud as an IT solution brings many benefit to high performance scientific computing. Moving forward, the hybrid cloud IT solution still has room for improvement.

The main problem is the cost of creating a private cloud locally. The hardware and software investment needed to create a private cloud on an organization's premise is quite prohibitive [107].

There is also administrative overhead of managing a private cloud as well as HPC clusters and virtual HPC nodes on the public cloud. Increased efficiency would further improve on the computing capabilities offered by the cloud service provider while helping them offer more competitive pricing, benefitting both service provider and end users. One of these areas is reducing the idling time of the public cloud, which can be used for heavy processing. The public cloud idling time is heavily influenced by the data transfer rate between the private and public clouds and the job scheduling algorithms. Current hardware and software infrastructure in the existing on premise HPC cluster would have to be reconfigured to accommodate the private cloud software and hardware infrastructure.

Improved and specialized algorithms for managing data transfer between the clouds can ensure that the public cloud need not idle long while waiting for essential data to be uploaded. Complementing well-designed data transfer algorithms with multiple channels of communication would further improve data transfer rates. As for job scheduling, ensuring that the jobs are scheduled in ways that minimize compute idling time and takes full advantage of parallel processing is also paramount for improving performance. Other ways to increase the performance of the cloud would be better software-hardware integration. Software that is specifically designed to fully utilize the hardware it is running on would boost performance tremendously. This is made easier as the data centers hosting the public clouds tend to have standardized hardware making it easier to create efficient software.

A major challenge with the hybrid HPC cloud computing model is that the system needs a way to advise the HPC cluster administrator of the best implementation for a particular application. For instance, HPC applications like ABAQUS and STAR CCM+ require mechanisms in place to analyse the nature of the application's computing load and make recommendations to the HPC cluster administrator on whether to keep the application on premise, off-premise or in the hybrid cloud processing [101].

Any hybrid cloud based approach to HPC would need to implement guidelines that will determine the division of processing load for applications based on three factors. Firstly, the system would need to determine application architecture as a Message Passing Interface or Service Orientated architecture [108]. Secondly, some applications are vendor-locked meaning that the vendor has predetermined whether the application is suitable for either on premise, off-premise or hybrid cloud processing [53]. Lastly, computational load analysis of all the current jobs on the HPC cluster needs to be done to determine if a running job needs more resources from the public cloud to complete faster.

Ideally, all computing jobs sent to the HPC cluster are monitored in real-time to determine if new jobs can be undertaken on premise or if the job needs to be sent fully off-premise or processed in the hybrid environment [101]. Computational

Load on the HPC cluster is dynamic, therefore changes in cluster load are monitored and new jobs are allocated to it in real-time.

The hybrid cloud should also aim to increase platform agnosticism to increase
usability. As every part of the IT infrastructure is virtualized, it is possible to create
interfaces between virtual machines that would allow end users to seamlessly
transfer data and computing jobs between Operating Systems. This can be realized
by integrating intermediary software in the cloud that can handle inter-OS communication and data transfers. The platform agnosticism concept is embodied in the
web-based control panel of the public cloud which can be accessed from any OS
with a web browser and should be extended to controlling the private cloud
component of the hybrid as well.

2.1.8   HPC + Cloud Contrasted with Existing Implementation Architectures

The key distinctive difference in the HPC + Cloud architectures is that it provides
on demand scaling at a relatively low startup cost, The low startup cost obviously
comes from the fact that it utilizes existing on premise HPC cluster processing
nodes and provisions more processing nodes from the cloud. It reduces high latency
issues faced by cloud technology by using hybrid cloud technology specifically the
application layer, transport layer, and session layer stack optimizations that enable
private cloud and public cloud to seamlessly synchronize.

HPC on Cloud seems at first glance a great solution as it also provides on
demand scaling, the cost of migrating a HPC cluster entirely onto the cloud is
prohibitively expensive, especially startup costs. This is because, existing hardware
and software that is physically on premise has to be virtually replicated entirely on
the cloud. Unlike the HPC + Cloud that only provisions extra processing nodes if
needed from the cloud and therefore has only part of it its infrastructure in the
cloud.

HPC on Hybrid cloud also seems to be a good solution as it provides on demand
scaling, however unlike HPC + Cloud which utilizes the existing local on premise
infrastructure. The HPC on hybrid cloud requires for there to be significant new
hardware and software investment in converting the current on premise infrastructure into a private cloud. Depending on requirements cost of maintaining a
private cloud is quite prohibitive.

HPC on Cloud, HPC on Hybrid Cloud and HPC on Grid do face some data
privacy issues. Some data is legally bound to stay within the local on premise
infrastructure. For example: medical data, student data and banking data.
HPC + Cloud overcomes this problem by having HPC + Cloud software framework that acts a gatekeeper and selectively chooses which process can migrate to
the cloud and which processes can only stay local on premise.

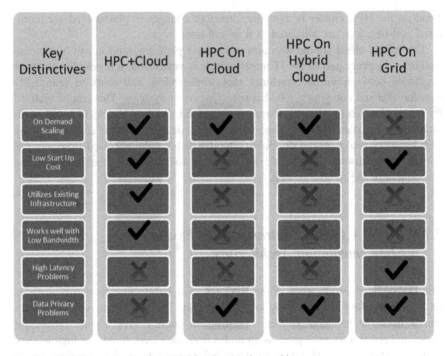

Fig. 2.7 Key distinctive of various HPC implementation architectures

HPC on grid on the grid does provide on demand scaling in a severely limited fashion as processes can be distributed to the many nodes on the grid cluster. This raises another major problem as grids are typically scattered geographically causing latency problems to affect the HPC processing time. Grid computing platform do not provide any guarantees of demand scaling and provide the services on a most optimal effort basis. Unlike HPC + Cloud, in which the cloud service provider service level provides agreements with guarantees of bandwidth and uptime (Fig. 2.7).

2.1.9 Profit-Based Scheduling

There is a business trend nowadays which utilizes cloud computing to reduce total cost of equipment ownership due to its cost-effectiveness, scalability of size and flexibility of management. On service user side, cost is simply based on the usage but for IaaS service provider, calculation of cost is relatively more complicated. Sharing of IaaS cloud resources between multiple service users is a crucial key behind the cost and profit management in the IaaS cloud [109] introduced an algorithm to maximize profit by using composite service while maintaining service quality such as response time which are specified by service users.

To process an application submitted by service user, a set of services s, $s = \{s_0, s_1 \ldots s_n\}$ to be created for it based on directed acyclic graph (DAG) and each of the services is interdependent on each other with precedence-constrained where it is applied. There are two sets of algorithm, first set of algorithms concerns about the profit achievable from current service and other services which both are running on the same service instance. This will result the highest profitable service run on same service instance while the other service will be transferred and run on another newly assigned service instance. Service instances will be dynamically assigned and fulfil the services respectively to achieve the target of composite service.

Second set of algorithm is as similar as the first set but the focusing key is different which is on the maximization of service instance utilization. Each service selects the lowest utilized instance to run the process. This is to maximize the profit based on improved resource utilization.

As the brief introduction above, it is able to realise several problems. The number of service instances are impossible to be unbounded due to the limitation of resource available. Assume that the number of service instances is large, in each of the service instance, the service is required to scan through every other service instance to match the rules of algorithm, and this will be a harsh environment with time complexity $O(n^2)$. All abovementioned problems are the reasons which these algorithms are not suitable for data intensive task because most of the service instances will be fully or almost fully utilized and it causes composite service relatively difficult to proceed. On the other hand, it is not optimal to process small data task because there will be only few service instances are utilized while the rest are idle without contribution. Keeping all service instances busy to process task can increase the profit gains.

2.1.10 Preemptable Scheduling

IaaS provider leases variety of computational resources to the service users indeed the preparation of handling service request simultaneously will be a must. Due to this reason, [110] introduced a resource optimization mechanism with preemptable function during every task execution.

After receiving the tasks from service users, estimate the average execution time of each task and arrange all tasks into a start time priority based list with ascending order. A matching process will be started by polling the resource status information such as earliest resource available time from all other service instances and assign the most optimal matched service instances for it. The tasks will then be distributed to the assigned service instance respectively. Tasks are defined in two types which are advanced reservation task with dominant priority in certain time is able to pre-empt all others which are called most optimal-effort task or zero reservation task during the matching process.

This phenomenon can be explained as overflow of workloads to another service instances which is able to enlarge the resource pool and provide even more flexible process because constant polling for necessary status of service instances. It may have a delayed but successful job done leads to a longer estimated finish time.

Service instances can be further defined as a member of federated heterogeneous multi-provider cloud system which each of the cloud systems is located at different geographical area. A manager server acts as the head of service instances in each cloud system to perform matching process for optimal resource allocation mechanism and the communication with other manager server. Head of the manager servers is a first contact point to receive tasks from service users.

Problems are found as expected as this is a decentralized approach which repeated crossing information between the manager servers will be a conflict to cause system collapse. As focusing on data intensive tasks, big data scheduling is not possible to produce real time result and the pre-emption of task is not a viable solution because the resource of computation will be mostly occupied during processing period. Lacks of memory to store the pre-empted task, excessive overheads of pre-emption, complicated recovery mechanism of pre-emption, continuous incoming processing tasks and lack of collision detection mechanism will cause entire cloud system become problematic and inaccessible.

2.1.11 Priority-Based Scheduling

In [111], large application can be decomposed and assigned into tasks with multiple sequences which can be represented by Directed Acyclic Graph (DAG). Each of the task will be weighted to get the estimated execution time and data transfer time between each task in the workflow.

Weighting Phase for this large application on DAG is considered completed and the algorithm enters prioritizing phase which compute the priority of the individual task by traversing the DAG upward while adding and assigning the accumulated priority value for both execution time and transfer time to the task from the ending point to starting point of DAG. For example, the time cost for Task A = 5, Task B = 6 and Task C = 7 and DAG created for these tasks is A → B → C. The priority value of A is 18, B is 13 and C is 7 according to the rule of prioritizing phase.

Mapping phase will be last phase to continue this algorithm, each task will choose the most optimal service instance that favours the scheduling process with pareto dominance formula which includes the calculations of various service instance parameter such as execution time, data transfer time, profit gain and a 0–1 floating point factor indicates the proportion of time and profit. Any idle service instance will be included in selection process to reduce slack time as well.

Dynamically mapping each tasks to the most-efficient service instance cost time complexity $O(n^2)$ which is not optimal algorithm especially in big data processing. Other disadvantages such as the DAG is application specific graph which is random

Table 2.2 Performance metrics for algorithms

Algorithms	Metrics				
	A	B	C	D	E
Profit-based scheduling [109]	High	High	Medium	High	Medium
Preemptable scheduling [110]	High	High	Medium	High	Low
Priority-based scheduling [111]	Medium	Medium	Medium	High	Low

A Reliability
B Ease of deployment
C Quality of service
D Delay
E Control overhead

in different situation. Random DAG is difficult in scalability when number of tasks is up to n numbers in unpredictable sequences. For example, it may require to overflow some tasks to other cloud platform during data intensive process and unknown sequences will cause a loss in data combination in finishing state and all effort will be in vain.

A brief performance comparisons has shown in Table 2.2.

2.2 Various Aspects and Current Issues

Task scheduling management is a challenging techniques on IaaS cloud to handle the fluctuations of workload submitted by service users at any random time while efficiently determine the demand of current limited demand of task to perform reliable quality of service guarantees. To achieve it, scheduling of tasks into a timetable of event with designed resource sharing in the according plan is affected by various aspects as specified in [112].

2.2.1 Service Provisioning

Provision of cloud resources to service users with load balancing mechanism and high available access time to optimize service quality while the operating cost of IaaS service is under control [113, 114]. Related issues are as follows.

- How to minimize the utilization of cloud resource provisioning to a task for increasing the profit gains but not breaking the agreements with service users?
- How to develop an optimal control mechanism on service instance to process the unpredictable incoming tasks with minimal loss?
- How to maximize the utilization of cloud resource to support multiple incoming task by providing actual needs?

- How to develop an algorithm for resource provisioning to be scalable up to multitier environment in IaaS cloud?
- How to develop a method to overcome bottleneck condition during task execution such as service instance is not responding?

2.2.2 Service Allocation

Allocation of proper cloud resources to proper task by using metric determination mechanism can reduce the operating cost and time especially in competing task situation such as pre-emption by higher priority task [115, 116]. Related issues are as follows.

- How to develop an algorithm to span the resource allocation between multiple clusters or data centres?
- How to develop a tree-based structure algorithm for resource allocation in IaaS cloud but still reserves the benefits of current algorithm?
- How to minimize the process overhead and performance degradation during resource allocation in cloud?
- How to design a user-oriented resource allocation mechanism with automated process management, risk management and customer service management without violating service level agreements?
- How to include the integration issues of multi-cloud providers?

2.2.3 Service Adaptation

Some parameters of task may be requested by service users to modify at any time so that adaptive control mechanism should be available for service users to fulfil their requirements [117, 118]. The reason why service users request is to maximize their benefits within the task running processes. Increase the control flexibility can ensure the service users get exactly what they require especially in high workload transitions and the outcome is out of their expectation. Related issues are as follows.

- How fast and frequent can the IaaS cloud respond to the service user requests?
- How complicated is the task submitted by service users as highly complicated task can be rejected by the IaaS service provider due to profit and hidden risk?
- How to design an algorithm for task restoration and recovery to the previous stable states if the error occurs?
- How to develop an automated recovery algorithm to bring the process to normal state?

- How to design a platform to let service user monitor the entire detailed processes thoroughly in real time but enclose the sensitive information in terms of security?

2.2.4 Service Mapping

Mapping of cloud physical resources to service instances has impact on service availability to service user because the concept of maximizing cloud utilization while minimizing physical resources procurement and maintenance is a target to achieve [119, 120]. Related issues are as follows.

- How to map service instances on physical resources wisely to meet the requirement based on the certain constraints of physical resource?
- How to develop an algorithm to map resources with least time consumed, cost and meet all deadlines of task?
- How to validate whether the task is suitable to process in the IaaS cloud before mapping process that can prevent unnecessary work done?
- How to design an algorithm to predict the performance of physical resources in stress situation?
- How to balance the load on physical resources while least affecting the process running on service instances that have been already mapped earlier?

Chapter 3
Benchmarking

Three public clouds was benchmarked which are Microsoft Azure, Amazon EC2 and the Australian National eResearch Collaboration Tools and Resources (NECTAR). NECTAR is an Australian Government project to provide public cloud resources to Australian Universities. The other player of choice among both Industry and Academic Institutions is Amazon Public Cloud or commonly known as Amazon EC2 which is a subsidiary of retail giant Amazon.com. The main reason the Amazon EC2 is popular with both academia and industry is because Amazon was an early pioneer in providing public cloud services. Compared to both NECTAR and Amazon Public Cloud, Microsoft Azure is a relative newcomer only starting to provide its services in 2012.

To determine which public cloud provided the most optimal high performance computing processing prowess. The software used to the benchmark the cloud was Roy Longbottom's Linux benchmarking tools [60].

3.1 Cloud Benchmarking Instance Specifications and Assumptions

For cloud computing instances (an instance is unit of computing resource provided by a cloud provider), the cloud provider provides a fixed computing instance specification meaning there is no way to adjust the CPU option and memory option to ensure parity between the different providers. The specifications above are all in the medium instance for each provider at the time of running the benchmark (Table 3.1).

© Springer International Publishing AG 2018
R. K. J. Tan et al., *Optimized Cloud Based Scheduling*, Studies in Computational Intelligence 759, https://doi.org/10.1007/978-3-319-73214-5_3

Table 3.1 Cloud benchmarking instance specifications

	Windows Azure	Amazon	NECTAR
Processor	AMD Opteron™ Processor 4171 HE	Intel® Xeon® CPU E5-26S0 0 @ 2.00 GHz	Intel® Core™ 2 Duo CPU T7700 @ 2.40 GHz
Measured	Minimum 2095 MHz, Maximum 2095 MHz	Minimum 1800 MHz, Maximum 1800 MHz	Minimum 2600 MHz, Maximum 2600 MHz
CPUs	2	2	2
RAM size (GB)	3.36	3.66	7.80

3.2 Classic Benchmark Test Categories

3.2.1 Dhrystone Benchmark

This test is a performance measurement of integer. There are two available versions, 1.1 and 2.1. The difference between 1.1 and 2.1 is that 2.1 avoid over-optimization problems that have been encountered in version 1. Table 3.2 shows a comparison between the three platforms the benchmark tests were running on namely, Amazon, NECTAR and Windows Azure. These tests include versions 1.1 and 2.1, with and without optimization. The ratings obtained are that of VAX MIPS where VAX stands for Virtual Address eXtension and MIPS means Million Instruction per Second.

3.2.2 Linpack Benchmark

This test measures the floating point computing power of a system. Floating point shows a way of representing the approximation of a real number in such a way that it can support a wide range of values. The Millions Floating-point Operations per Second (MFLOPS) is the unit by which the benchmark test is measured.

Table 3.2 Classic benchmark test results (higher value denotes better performance)

	Windows Azure	Amazon	NECTAR
Dhrystone Benchmark (VAX MIPS rating)	8155	10,455.50	10,752.28
Linpack Double Precision Unrolled Benchmark (MFLOPS)	1317.95	1603.02	1609.31
Livermore Loops Benchmark Maximum Rating (MFLOPS)	2588.9	2733.8	2634.1
Whetstone Single Precision C Benchmark MWIPS (MFLOPS)	2135.854	2111.706	2644.834

3.2.3 Livermore Loops

Livermore loops is a benchmark test that is usually running for parallel computers. Produced for supercomputers comprising of numerous kernels, three specific types of data sizes are ran and the results obtained are in MFLOPS.

The results generated for overall ratings consist of Maximum, Average, Geometric mean (Geomean), Harmonic mean (Harmean) and Minimum whereby Geomean is the official overall rating. All tests for Livermore loops were completed over 24 loops and the geometric mean was the one recorded.

3.2.4 Whetstone Benchmark

The Whetstone Single Precision C Benchmark is related to CPU performance and is meant to check speed ratings in Millions of Whetstone Instructions per Second (MWIPS).

In the Dhrystone Benchmark Performance NECTAR scores $1.3\times$ better than Windows Azure however between NECTAR and Amazon, NECTAR scores $1.03\times$ better. NECTAR is the most optimal for this benchmark. In the Linpack Benchmark Performance NECTAR scores $1.22\times$ better than Windows Azure however between NECTAR and Amazon, NECTAR scores $1.004\times$ better. NECTAR is the most optimal for this benchmark. In the Livermore Loops Benchmark Performance, Amazon scores $1.056\times$ better than Windows Azure and between Amazon and NECTAR, Amazon scores $1.038\times$ better. Amazon is the most optimal for this benchmark. In the Whetstone Single Precision C Benchmark Performance, NECTAR scores $1.24\times$ better than Windows Azure however between NECTAR and Amazon, NECTAR scores $1.25\times$ better. NECTAR is the most optimal for this benchmark. Overall, NECTAR is the most optimal in this classic benchmarks category followed by Amazon and Windows Azure (Fig. 3.1).

3.3 Disk, USB and LAN Benchmarks

This test makes use of direct Input Output (I/O) for the speed of Local Area Network (LAN) to avoid data from being cached in the main memory of the Operating System. Also involved in the benchmark tests are the read/write speed. In this test, a 64 kb file was written, read and deleted 500 times and the result can be seen in Table 3.3.

Disk, USB and LAN performed is critical in determining the processing throughput of a high performance computing cluster, as no matter how fast the CPU is, final processing times are constrained by I/O operations that are ultimately determined by the read and write speed of the Disc, USB and Local Area Network

Fig. 3.1 Classic benchmark results (higher value denotes better performance)

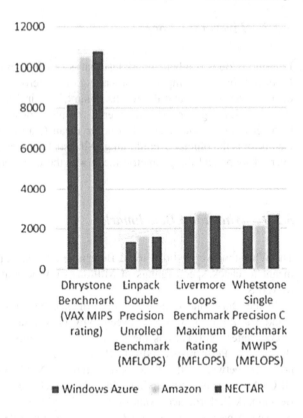

■ Windows Azure   Amazon   ■ NECTAR

Table 3.3 Disk, USB and LAN benchmark test results

	Windows Azure	Amazon	NECTAR
Write MB/s	122.83	25.34	6.23
Read MB/s	274.74	67.75	92.84

(LAN) interfaces of the processing node. In the Disk, USB and LAN Benchmarks, for the write category, Windows Azure is the most optimal followed by Amazon and NECTAR while for read category, Windows Azure is still the most optimal followed by NECTAR and Amazon. Overall, Windows Azure is the most optimal followed by Amazon and NECTAR (Fig. 3.2).

3.4 Multithreading Benchmarks

Multithreading influences high performance computing as it shows the efficiency at which a high performance a computer manages multiple concurrent processes.

Fig. 3.2 Disk, USB and LAN benchmarks (higher value denotes better performance)

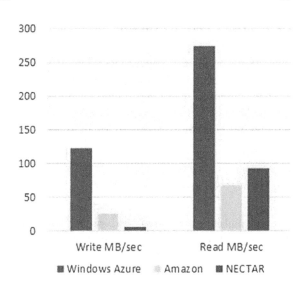

3.4.1 Simple Add Tests

The tests involved in Simple Add Tests execute 32 bit and 64 bit integer instructions as well as 128 bit SSE floating point. The performance is very relative to the amount of CPU cores available in the system. Since the benchmark test is about multithreading, each thread is given an independent code to test for each thread. The values taken for this test is the average of two aggregates tested individually.

3.4.2 Whetstone Benchmark

As opposed to the previous whetstone benchmark, this test focuses on multithreading application. Again, the number of cores present is a determinant factor on the speed of the test run. The results taken as reference for the test are based on the time taken for the last thread to finish and measured in Millions of Whetstone Instructions per Second (MWIPS).

3.4.3 MP FLOPS Program

The purpose of this test is to check for the multiplication of floating point calculations with data from higher level of caches or RAM. These programs can be used as a burn-in/reliability test and similar functions can be run on a different segment of data. The last Million Floating Point Instructions per Second (MFLOPS) value from the test is taken as reference.

3.4.4 MP Memory Speed Tests

This test makes use of single and double precision floating point numbers and integers to test for the speed of the memory. The average value of the read, write and delete were taken individually and then graphed to figure out the most optimal out of the cloud systems.

3.4.5 MP Memory Bus Speed Tests

The bus/memory speed is tested by reading all the data at the same time. The value taken for this test is the ReadAll value of the largest file. This accounts for a sizeable cache and RAM usage stressing the bus and allowing for an estimation of maximum bus/memory speed.

3.4.6 MP Memory Random Access Speed Benchmark

This benchmark test is about read and read/write tests that cover cache and RAM data sizes. The largest file (96 MB) is chosen since it uses all the resources and maximises the stress on the cores for the test giving a very relatable value. The average of the serial read, read/write and random read, read/write as well as mutex read, read/write is taken to give a general idea of how it performs on various platforms (Table 3.4).

For multithreading benchmarks, NECTAR is the most optimal is in 5 out of 6 categories while Amazon is the most optimal in 1 out of 6 categories. Windows Azure fared badly in all categories however in the multithreading add test and multithreading double precision whetstones, it was close to NECTAR and the overall most optimal performer in multithreading is NECTAR followed by Amazon and Windows Azure (Fig. 3.3).

Table 3.4 Multithreading benchmarking test results (higher value denotes better performance)

	Windows Azure	Amazon	NECTAR
Bus speed (MB/s)	5582	7574	4961
MP memory reading speed (MB/s)	5902.44	6183.44	6715.44
Multithreading add test (Million instruction per second)	13,874	6343	14,552.50
MP SSE MFLOPS benchmark (MFLOPS)	18,418	10,018	27,839
Multithreading double precision whetstones (MWIPS)	4257	1981	5012
RandMemMP speeds (Mbytes per second)	2372.33	2608.33	2622.67

Fig. 3.3 Multithreading benchmarking (higher value denotes better performance)

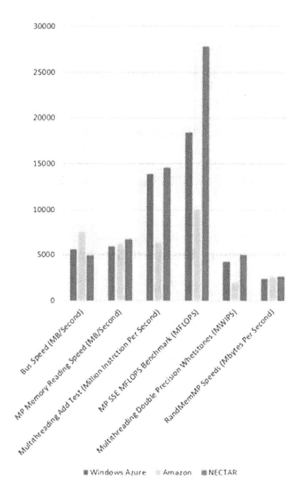

3.5 OpenMP Benchmarks for Parallel Processing Performance

3.5.1 MemSpeed

This test makes use of single and double precision floating point numbers and integers to test for the speed of the memory. The average value of the read, write and delete were taken individually and then graphed to figure out the most optimal out of the cloud systems.

3.5.2 Original Open MP Benchmark

Taking the MFLOPS value, this test behaves in a similar way as Windows compilation, meaning the performance gains of the number of cores present is relative to the time taken for the test to complete as compared to a single core. The average value for data in and out is taken as the comparison value for the different platforms (Table 3.5).

For the OpenMP benchmarks, Amazon is the most optimal in memory reading speed test followed by Windows Azure and NECTAR while for OpenMP MFLOPS benchmark, NECTAR is the most optimal followed by Amazon and Windows Azure. Overall, NECTAR and Amazon is the most optimal followed by Windows Azure (Fig. 3.4).

3.6 Memory BusSpeed Benchmark

3.6.1 BusSpeed Test

This test makes use of single and double precision floating point numbers and integers to test for the speed of the memory. The average value of the read, write and delete were taken individually and then graphed to figure out the most optimal out of the cloud systems.

Table 3.5 OpenMP benchmarking test results (higher value denotes better performance)

	Windows Azure	Amazon	NECTAR
Memory reading speed test (MB/s)	4147.33	6457.78	4095
OpenMP MFLOPS benchmark (MFLOPS)	4781	10,035	13,886

Fig. 3.4 OpenMP benchmarking (higher value denotes better performance)

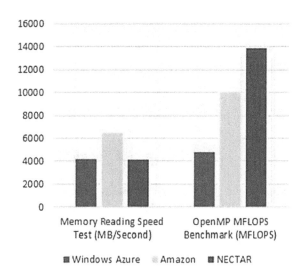

3.6.2 Random Memory Benchmark

This test shows the behaviour of the memory with increasing file size in terms of data transfer. The values taken are similar to that of the MP Memory tests.

3.6.3 SSE Benchmark

This variation of the previous SSE benchmark excludes AMD but measures Single Precision and Double Precision, floating point speeds, data streaming from caches and RAM. The alterations in this test avoid intermediate register to register operations to produce much faster speeds. Again, the largest value is taken as reference and compared across platforms (Table 3.6).

Table 3.6 Memory BusSpeed benchmark results (higher value denotes better performance)

	Windows Azure	Amazon	NECTAR
Bus speed test (MB/s)	5455	7461	2525
Random/serial memory test (MB/s)	1850.25	3113.625	1910.375
SSE & SSE2 memory reading speed test (MFLOPS)	4935.25	4267.25	5956.125

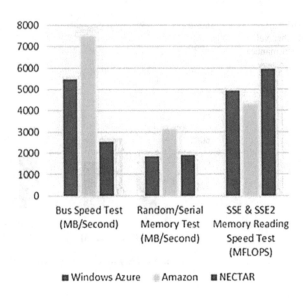

Fig. 3.5 Memory BusSpeed benchmark results (higher value denotes better performance)

Table 3.7 Overall results

	Windows Azure	Amazon	NECTAR
Classic benchmarks for CPU performance	Third	Second	First
Disk, USB and LAN benchmarks	First	Second	Third
Multithreading benchmarks	Third	Second	First
OpenMP benchmark for parallel processing performance	Third	Second	First
Memory BusSpeed benchmark	Third	First	Second

In BusSpeed benchmark, Amazon is the most optimal in bus speed test followed by Windows Azure and NECTAR while in random/serial memory test, Amazon is still the most optimal followed by NECTAR and Windows Azure. For SSE and SSE2 memory reading speed test, NECTAR is the most optimal followed by Windows Azure and Amazon. Overall in this BusSpeed benchmark, Amazon is the most optimal followed by NECTAR and Windows Azure.

The main aim of this benchmarking is to identify the most optimal possible candidates for a building HPC + Cloud high performance cluster that can be deployed quickly and easily. Windows Azure came in last in almost all categories.

However given that HPC + Cloud will require high I/O throughput between on premise HPC which is currently in the organization and the public cloud, the high I/O throughput is necessary to avoid data bottlenecks between the HPC and the Cloud.

Therefore Windows Azure, despite coming in last place in all categories except Disk USB and LAN benchmarks still is the prime candidate for deployment for the HPC + Cloud (Fig. 3.5 and Table 3.7).

Chapter 4
Computation of Large Datasets

With our solution, existing IT solutions of enterprises can be integrated with the cloud hence increasing the breadth of applications supported and improving performance with reduced additional overhead since additional storage and computing power become easily available on demand. This capability will be demonstrated by the prototype created at the end of the project.

Enterprises that create an IT solution using the prototype can reduce the large up-front investment in expensive equipment since computational power can be utilized from the cloud reducing required on-site IT assets. Enterprises can acquire a cost effective cloud management system for their hybrid cloud solution. Increasing automation enables enterprises to avoid costly consultant help when migrating to the hybrid cloud environment. Simpler and more user-friendly interfaces further accentuates the lower manpower requirements of the cloud environment.

From a technical point of view, this project will provide a method to migrate to and apply a hybrid cloud environment for enterprise that offers improved processing in science and engineering applications. The environment will be able to intelligently scale available resources, onsite or from the public cloud, and observe restrictions when running applications. Another feature of the environment will be the ability to intelligently monitor system status and handle errors. Proof of concepts will be created to demonstrate the capabilities and advantages of the applied hybrid cloud environment which will also serve as case studies for that can be applied when integrating other applications.

© Springer International Publishing AG 2018
R. K. J. Tan et al., *Optimized Cloud Based Scheduling*, Studies in Computational
Intelligence 759, https://doi.org/10.1007/978-3-319-73214-5_4

4.1   Challenges and Considerations in Computing Large-Scale Data

Large scale computing that can reach exascale computing will provide the computational power essential for addressing many valuable scientific challenges. However, despite the performance of our current technologies that have achieved petascale, scaling up that technology another order of magnitude to exascale is not possible. Numerous reports have been produced by the Department of Energy in the United States that document the technical challenges and the non-viability of the existing computer designs to reach exascale [121–123]. Those challenges include but are not limited to the followings.

4.1.1   Extreme Parallelism and Heterogeneity

The rate at which processor power increases is predicted by Moore's law but as we approach the limitations in our manufacturing processes and limits in component power density, processor clock speeds have stagnated. To continue increasing the performance of our computing devices, the number of processing elements on a chip (multiple cores) and multi-threading support was implemented. It is estimated that exascale machines will have two to three orders of magnitude of parallelism over petascale computer levels, which is much greater parallelism on nodes than is available today. The bulk-synchronous execution models that dominate today's parallel applications will not scale to this level of parallelism.

New algorithms need to be developed that identify and leverage more concurrency and that reduce synchronization and communication. One approach will be through dynamically scheduled task parallelism; but this will introduce a new challenge, reproducibility, that will make determinations of code correctness more difficult. This problem would be further exacerbated if multiple systems are employed simultaneously with heterogeneous software and architectures. Though combining multiple systems to meet a computing goal is an advancement upon the multi-core idea, it will introduce even further complexity to task management and scheduling algorithms.

4.1.2   Data Transfer Costs

Another potential bottleneck is that memory bandwidth is not expected to increase at the same rate as the number of processing units. Consequently, even though overall, on-nod, memory bandwidth will increase, the bandwidth per core will actually decrease. Interconnect transfer rates are also not expected to increase at the same rate as the number of cores. Additionally, the energy used for a double-precision flop is expected to decrease by roughly an order of magnitude, which will further expose the energy cost for off-chip data motion and on-chip transfers as the largest energy sink in data processing. In the future, a variety of different memory technologies may be employed to alleviate this issue including nonvolatile memory, stacked memory, scratchpad memory, processor-in-memory, and deep cache hierarchies.

If further interconnected computing resources are to be employed, algorithms will also need to be more aware of data locality and seek to minimize data motion, since this will become a more significant energy cost than computation.

4.1.3 Increased Failure Rates

Because of the staggering number of components expected on exascale computing systems, hardware failures are expected to increase. Traditional error correction methods, such as the checkpoint-restart recovery mechanism, become too expensive in terms of time and energy due to bulk synchronization and I/O with the file systems at exascale. An exascale system that employs global recoveries could conceivably take more time to ready itself than the mean time between failures.

Local recovery mechanisms that are tailored to the tasks being run to minimize system down time upon error detection are required. Further errors are to be expected as efforts to reduce power consumption by computing at lower threshold voltages and other environmental disturbances may lead to more soft errors that may not be caught by hardware error detection mechanisms. Increased management software complexity as the system expands also increases the likelihood of errors. These increased fault rates will affect all hardware in the stack, hence, applications will need to be fault-aware and employ algorithms to make them tolerant to certain types of faults. If not dealt with properly, the sum of all the errors introduced by failures and system recoveries will make it difficult to reproduce computations, which will lead to difficulties in assessing the effectiveness of any code being tested.

4.1.4 Power Requirements at Exascale

Power consumption is the new primary motive for changes in supercomputer architecture. When implemented, exascale computing would have to be more focused as "low-power, high-performance computing." Scaling up our current systems to the exascale is not viable since the power requirements of such a machine rapidly become prohibitive [124]. Reports done for the United States' Department of Energy by Dongarra et al. has therefore set a target of achieving exaflop performance with a power limit of 20 MW [125]. This restriction has direct implications for the structure and organization of hardware components in any exascale system as well as algorithms employed. In the future, energy used by a computational task may become the new cost metric that replaces the current metric of CPU time. This implies that numerical algorithms will need to become more power-aware.

4.1.5 Memory Requirements at Exascale

Without a significant change in technology, memory density is also not expected to increase at the same rate as the number of processing units. For memory, power is a major limiting factor. Our current systems employ volatile RAM technology which consumes a great deal of power to maintain its state. Therefore, while the amount of memory per node will continue to increase, the difference in rate of increase means that the amount of memory per core will still decrease. Algorithms that function well now may become memory constrained when scaled to an exascale system employing our current memory technologies. These algorithms will need to be redesigned to minimize memory usage.

The solutions to the above challenges would introduce key architectural changes which will be necessary to build an exascale machine. The implementation of those changes will affect the software stack in ways that will affect applications and numerical algorithms run on them in turn. Particular constraints that need to be considered when considering an exascale system include the presence of distinct computing/architectural layers, multiple levels of parallelism; the high cost of data motion across architectural layers; the lack of efficient resiliency when dealing with both soft and hard errors and the power and memory resource consumption of a system of that scale.

4.2 Work Done to Enable Computation of Large Datasets

It has become obvious that many changes will have to be made to compute large data sets that will take advantage of the technology we currently have and incorporate future advances in computing technology. The most suitable solution currently available would be to take advantage of cloud technology supplementing our current traditional High Performance Computing clusters. Another consideration is that at large scales, efficiency of the tasks become very application specific meaning that it is possible to achieve exascale level computing with our current technologies with a few caveats.

Create an environment that integrates existing high performance computing clusters at universities with external cloud computing infrastructures so that the computational resources can be increased when they are needed by the applications and are released when they are not in use. This hybrid model can be adapted to enterprises, allowing them to decrease their annual ICT infrastructure costs while maintaining the efficiency and scalability of the applications that run on it. It differs from other available clouds by automatically implementing task management and networking to multiple disparate computing resources so that it can be used as a singular, scalable, cloud computing resource. This is tested through an architectural case study where multiple clouds are connected together and tasks are run on the resulting architecture.

4.2.1 OpenStack and Azure Implementation

The first step taken was to establish multiple cloud environments to ensure inter-cloud connectivity. The two clouds implemented were OpenStack and Azure Pack which creates private clouds which would then be connected to the Microsoft Azure public cloud.

4.2.1.1 Introduction to OpenStack

OpenStack is an open source cloud computing platform that can support all types of cloud environments. OpenStack offers the same open source cloud solution as you would find on Eucalyptus, and also manages to outdo Eucalyptus when it comes to support and troubleshooting. OpenStack offers PaaS, IaaS and even NaaS in its newer releases.

Besides OpenStack offering a cloud computing software solution, it also contains many additional features and a number of tools [126]. These include: Scaling in size depending on demand and user needs; and processing big data and heavy workloads with tools like Hadoop High-performance Computing (HPC) environments. OpenStack has deployed its platform with PaaS and IaaS concepts in mind and supports a wide variety of hardware including the ARM processor architecture. It transcends both services and manages to tie in neatly with the packages from other clouds, such as Windows Azure [128]. Because of its universal compatibility with most high level programming languages, it can be considered as an IaaS service model. The cloud frame work implemented is shown in Fig. 4.1.

The open source cloud computing frame work consists of three important parts as discussed below:

- **Compute Node**: Offers on-demand computing resources by provisioning and managing large networks of virtual machines. Compute resources are accessible via APIs for developers building cloud applications and via web interfaces for end users.
- **Storage Node**: Offers object storage and block storage, with many deployment options depending on use case.
- **Network Node**: Offers pluggable, scalable and an API-driven system for managing networks and IP addresses. It can also be used to increase the value of existing datacenter assets. Lastly, it also ensures the network will not bottleneck

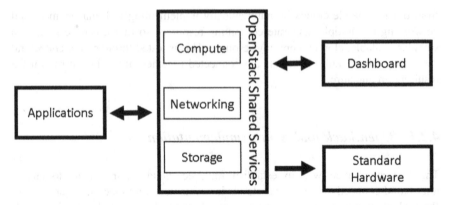

Fig. 4.1 Open source cloud computing framework

a cloud deployment and gives end users real self-service, even over their net-work configurations [126].

4.2.1.2 Network Architecture of OpenStack

The OpenStack implementation consisted of a three basic node architecture as shown in Fig. 4.2. This architecture provides high end computing, networking and storage facilities.

In Fig. 4.2 the first block section is the Controller node, the second is the Network node, and the third is for a Compute node. Additional Compute nodes can be added to the architecture to expand the clouds limits.

The controller node is responsible for the following basic services: identity, image, management portion of Compute and Networking, Networking plug-in and the dashboard. It also runs additional supporting features such as a message broker, database in MySQL and Network time Protocol (NTP) [126].

The Network node executes the following services: networking ML2 plugin, layer 2 and layer 3 agents that provide and operate tenant networks. The main role of layer 3 is routing, network address translation and DHCP services. Providing virtual networks and tunnels is taken care of by layer 2. Compute nodes execute the hypervisor part of compute services that functions as tenant virtual machines or instances. It also runs networking plug-ins and other optional services.

The OpenStack community has collaboratively identified nine key components that are part of the "core" of OpenStack, which are distributed as a part of any OpenStack system and maintained by the OpenStack community. By provisioning and managing large networks of virtual machines, these components enable enterprises and service providers to offer on-demand computing resources. For this implementation, only 6 of the components were necessary, as listed below:

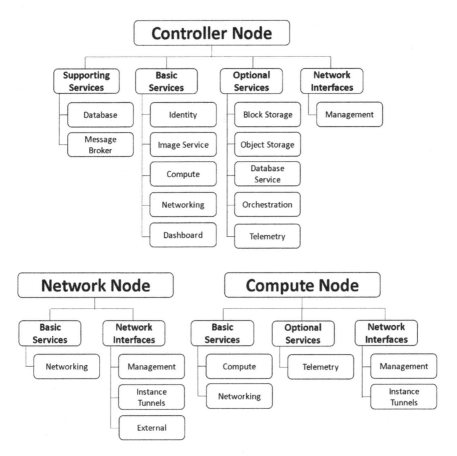

Fig. 4.2 Architecture of three nodes in an open source cloud

(a) **Basic Services: Identity Service (Keystone)**

Keystone provides an authentication and authorization service for other OpenStack services. Keystone integrates with LDAP to provide a central list of all users of the OpenStack cloud and allows administrators to set policies that control which resources various users have access to. It provides multiple means of access, meaning developers can easily map their existing user access methods against Keystone.

(b) **Basic Services: Dashboard Service (Horizon)**

Horizon provides administrators and users a graphical interface to access, provision, and automate cloud-based resources. Developers can access all of the components of OpenStack individually through an application programming interface (API), but the dashboard provides system administrators a look at what is going on in the cloud, and to manage it as needed. It's the primary way for accessing resources if API calls are not used.

(c) **Basic Services: Image Service (Glance)**

Glance stores and retrieves virtual machine disk images. It allows these images to be used as templates when deploying new virtual machine instances. One of the main benefits to a cloud platform is the ability to spin up virtual machines when users request them. By creating templates for virtual machines, Glance helps to achieve this benefit. Also, it can copy or snapshot a virtual machine image and later on allow it to be recreated. Glance can also be used to back up existing images to save them. Glance integrates with Cinder to store the images. OpenStack Compute makes use of these stored images during instance provisioning.

(d) **Basic Services: Networking (Neutron)**

Neutron (formerly Quantum) provides the networking capability for OpenStack. It helps to ensure that each of the components of an OpenStack deployment can communicate with one another quickly and efficiently. Neutron manages the networking associated with OpenStack clouds. It is an API-driven system that allows administrators or users to customize network settings. It supports the Open Flow software defined networking protocol and plugins are available for services such as intrusion detection, load balancing and firewalls.

(e) **Basic Services: Compute Service (Nova)**

Nova is designed to manage and automate the provisioning of compute resources. This is the core of the virtual machine management software, but it is not a hypervisor. Instead, Nova supports virtualization technologies including KVM, Xen, ESX and Hyper-V, and it can run on bare-metal and high performance computing configurations too. Compute resources are available via APIs for developers and through web interfaces for administrators and users. The compute architecture is designed to scale horizontally on standard hardware.

(f) **Basic Services: Database Service (Trove)**

Trove is the database as a service open source project for OpenStack. It is to provide scalable and reliable cloud database as a service, provisioning functionality for relational and non-relational database engines, and to improve its full-featured and extensible open source framework. Trove is designed to run entirely on OpenStack. Cloud users and database administrators can provision and manage multiple database instances as needed. Initially, the service will focus on providing resource isolation at high performance while automating complex administrative tasks including deployment, configuration, patching, backups, restores and monitoring.

4.2.1.3   Windows Azure Pack

The second cloud created was another private cloud implemented through Windows Azure Pack (WAP). WAP is a collection of Microsoft Azure technologies available to Microsoft customers at no additional cost. It integrates with Windows Server, System Center, and SQL Server to offer a self-service portal and cloud services such as virtual machine hosting (IaaS), database as a services (DBaaS), scalable web app hosting (PaaS), and more.

WAP provides a link between private and public clouds by providing a consistent experience between Windows Azure and private clouds. WAP allows a private cloud hosted on a local data center to replicate Windows Azure like capabilities. WAP leverages the same console and API technology used in Azure, and this brings a consistent platform of portal + API between private, public, and hosted clouds. WAP offers a series of services to its consumers by providing an Azure-like experience that includes a consistent interface, as well as a common API, that enables a consistent way to consume these services. These services include, but are not limited to, IaaS, Web PaaS and Database as a Service (DBaaS). The architecture provided by WAP is summarized in Fig. 4.3 (referenced from WAP manual [128]).

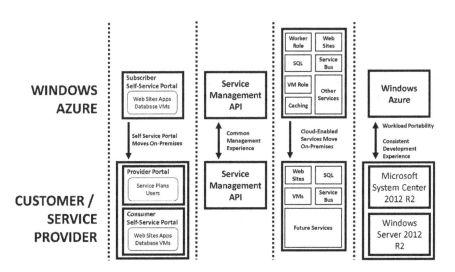

Fig. 4.3 Windows Azure Pack (WAP) architecture

4.2.1.4 Network Architecture of Windows Azure Pack

WAP architecture is an amalgamation of different web services which, when combined, offers an array of service layers.

There are two portals that makes up the WAP solution:

- **Admin Portal**: This portal lets you configure the different services offered via WAP, as well as defining plans (what services can be consumed and how much), and mapping these to subscriptions (who can consume services) so tenants can start using those services via the customer portal. The admin portal also offers the possibility to manage automation and metering for services consumed by customers. The admin portal resides inside the datacenter as an interface for WAP administrators.
- **Customer (tenant) Portal**: This portal allows tenants to consume services from WAP. These services include IaaS, Web PaaS, and DBaaS—and it enables 3rd party extensions to be used by customers. Using the WAP Tenant Portal, customers can manage these services in a way that's very similar to how services are managed in Azure. The WAP portal experience is almost identical and offers very similar capabilities for the service listed above as in Azure.

4.2.2 Cloud Interconnectivity

With the OpenStack cloud and Windows Azure Pack set up, connection between the clouds needed to be established to prepare the underlying infrastructure to be used to run tasks. Further connections were made to Microsoft Azure public cloud to create a single computing cloud that could run tasks.

To create the hybrid cloud environment, a connection between the virtual network and gateway in Azure Pack had to be established. This was done by first creating both the virtual network and the gateway and then retrieving a system-generated gateway IP address. Security for the established connection and during configuration of the VPN was maintained by generating a shared security key.

The VPN device was then configured with the appropriate network information and settings before a connection was established and traffic could move between the clouds using the internet as the platform for the hybrid cloud. All successful connections could be seen on the Azure Manage Portal status page. Data was successfully transmitted through the tunnel both ways proving a successful connection had indeed been established.

The overall Virtual Network topology established is shown in Fig. 4.4. The private cloud running Azure Pack is named Azure Virtual Network and the On Premises Network is the second, OpenStack private cloud running on Curtin's internal network. The Host is the Openstack Controller Node running on the Biomap domain with the computer's network name being BIOMAPHEADNODE.

The detailed topological breakdown for the On Premises Network is shown in Fig. 4.5. The firewall had a public IP address (113.23.137.110) with elevated

Fig. 4.4 Virtual network topology

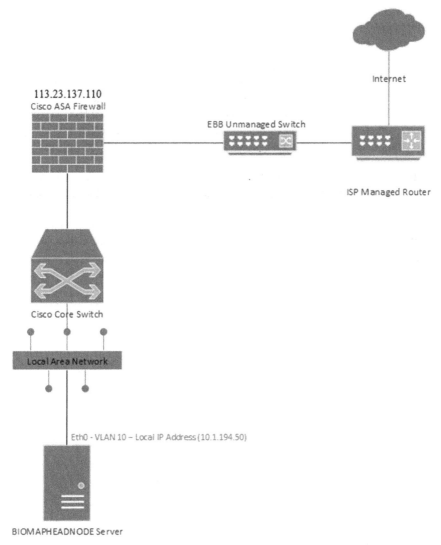

Fig. 4.5 On premises network detailed topology

security access that had been provisioned earlier for building the hybrid cloud environment. The internal IP address (Private IP) of the gateway to the OpenStack Cloud's network was 10.1.194.5. The Cisco ASA firewall also served as the host for the internal VPN that combined all the interconnected clouds. To connect another resource or cloud to the hybrid cloud environment, a user must first configure the new cloud to connect to the public IP of the network. Ultimately, a successful connection should be established to the primary controller node on OpenStack which acted as a Host with the IP address 10.1.194.50, and was named BIOMAPHEADNODE on the network.

4.2.3 HPC + Cloud

After the underlying infrastructure has been established, tasks sent to the hybrid cloud will be handled by a task management algorithm called HPC + Cloud. The main goal of the HPC + Cloud Software framework is to automate the sending of HPC applications jobs to the cloud when the resource on local HPC resources exceed a prefixed utilization threshold. Also the framework selectively chooses which processes can be migrated to cloud and if it is scalable according to the preconditions stated above (Fig. 4.6).

HPC + Cloud is the module that enables the HPC to link up with the Public Cloud. The HPC + Cloud consists of: Workload Monitoring (A) which monitors the HPC cluster utilization threshold. When the utilization threshold is reached by the HPC cluster, HPC jobs are sent to Batching (B) component which analyses the suitability of the job, according the stated preconditions of process data privacy, legal, and vendor limitations and also is one of these three types of processes: Message Passing Interface (MPI) type, service-oriented architecture (SOA) type,

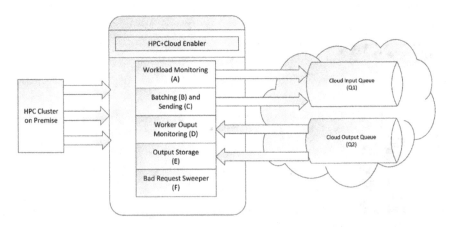

Fig. 4.6 HPC + Cloud framework

parametric sweep type or spreadsheet type. The Batching component (B) then updates the suitability of the process (Sj). The Sending part of (B) sends the suitable processes to the Cloud input Queue for processing in the cloud.

Upon process job completion, process jobs are place on Cloud Output Queue, Worker Output monitoring component (D) continuously monitors the Cloud Output Queue (C-out). Once there are completed jobs in the Cloud Output Queue. The Worker Output Monitoring component (D) triggers the Output Storage Component (E) to copy and store the data generated by the job in the local storage facility. In background continuous all the time the Bad Request Timeout Sweeper (F) runs continuously during steps 1–4. It functions to inform the HPC Head node of jobs that are not sent to the Cloud due to the jobs not meeting the preconditions for migration.

4.2.4   Simulations and Tools for Large Scale Data Processing

4.2.4.1   Performance of the Hybrid Cloud

To gauge the improvement in performance that the new computing environment could offer, simulations comparing a private cloud environment against the hybrid cloud environment were done. The simulator was built in-house and based on CloudSim [129]. The simulator was verified using tasks from Ray [130] run on multiple nodes for both Curtin University's HPC Lab Cluster and Microsoft Azure, representing the private and public cloud environment respectively. Ray is a genome assembler that generates tasks that are computationally intensive while maintaining support for parallel processing both on local clusters and the cloud.

Each simulation run consisted of 500 tasks with task parameter ranges taken from the Ray tasks run on both the private and public cloud environment in the verification stage. The same task queue was applied to both private and hybrid cloud environments and the estimated task completion time for both were calculated and compared. The characteristics of the tasks, local HPC cluster nodes, and the nodes in the public cloud are shown in Tables 4.1, 4.2 and 4.3.

A simulation of 500 tasks were chosen to ensure that the number of tasks exceeded the number of available nodes in the local cluster forming a queue. Performance enhancements provided by the hybrid environment are only noticeable when the local cluster is fully occupied and has jobs in queue. The tasks had mixed instruction lengths between 9 and 15 million instructions long to represent the

Table 4.1 Tasks details

Number of tasks	500
Instruction length (million instructions)	9–15
File size (GB)	0.1–100

Table 4.2 Local cluster node specifications

No. of cluster nodes	50
Processing power of each node (MIPS)	64,000
Hard disk space (GB)	250
RAM (GB)	32

Table 4.3 Public cloud specifications

Max number of requested nodes	25
Processing power of each node (MIPS)	64,000
Hard disk space (GB)	250
RAM (GB)	28
Bandwidth between local cluster and cloud (Gbps)	1
Maximum file size for processing on cloud (GB)	50

number of instructions in the more complex tasks. The file size is the size of associated input data and necessary data or library files that needed to be uploaded to perform the task on the cloud, and ranged from 0.1 to 100 GB.

Local nodes in the simulation were modelled after actual nodes in Curtin University's HPC Lab Cluster which are running 2.4 GHz Intel Xeon processors. The simulated public cloud nodes used the specifications of Microsoft Azure's Dv2-series instances. Each node contains a 2.4 GHz Intel Xeon E5-2673 v3 (Haswell) processor. Dv2-series nodes are specialized cloud nodes which are better suited for high performance computing applications that demand faster CPUs, better local disk performance, or higher memories.

An algorithm was applied that would send tasks in the local HPC cluster's job queue up to the public cloud when all 100 local nodes were occupied. Cloud resources would be requisitioned till a preset limit of 25 public cloud nodes was reached. Latency was offset by setting a file size limit for jobs sent to the public cloud. Any tasks that had file sizes exceeding 50 GB were returned to the queue for local processing. Relatively large files were not processed on the cloud to minimize file transfer time, which would have affected the overall processing time.

The completion time for each run for 30 runs of the 500 task simulation are recorded in Fig. 4.7. Under the simulation conditions, the average time for the simulated local cluster to complete 500 tasks was 352.4 h whereas the hybrid cloud environment took an average 245.9 h. Hence, an average performance improvement of 30.2% was obtained using the hybrid environment.

The results of the simulations shows that applying a hybrid cloud environment to Curtin University's HPC Lab Cluster would have resulted in sizable reductions in computation time for the users doing research there. This result can be scaled upwards to larger clusters where a total hybrid environment consisting of an additional 25% public cloud capacity would offer an approximately 30% increase in performance compared to just utilizing the local component alone. The overall better performance when utilizing the public cloud makes the hybrid the superior solution for computational needs in research institutes.

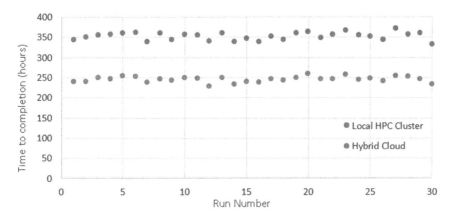

Fig. 4.7 Completion time of 500 tasks for local cluster and hybrid environment

The simulation showed that applying the hybrid cloud offers significant improvements in performance. The hybrid cloud offers the myriad advantages of the public cloud coupled with the specialized nature of the private cloud. Scaling up computing resources in a hybrid cloud cluster is much easier due to the flexibility of the public cloud and the increased scalability leads to cost savings and performance enhancements.

4.2.4.2   Further Simulation Work

Further simulation work is to be done after fine-tuning the simulation to be more inclusive of errors. The tasks that will be simulated will be from genome sequence assemblers, namely Ray and SOAP.

(a) **Ray**

Ray is a parallel software that computes de novo genome assemblies with next-generation sequencing data [130]. De novo genome assembly and taxonomic profiling are computationally intensive tasks which are normally run on distributed computing to handle the voluminous parallel sequencing datasets, especially those generated by metagenomic experiments, used as inputs. Ray profiles microbiomes based on uniquely-colored k-mers and is composed of Ray Meta, which is a massively distributed metagenome assembler, and Ray Communities, which utilizes bacterial genomes to color the assembled de Bruijn graph to obtain the complete genome. Ray is written in C++ and can run in parallel on numerous interconnected computers using the message-passing interface (MPI) standard.

Ray can accurately assemble and profile a three billion read metagenomic experiment representing 1000 bacterial genomes of uneven proportions in 15 h with 1024 processor cores, using only 1.5 GB per core. The software will facilitate the processing of large and complex datasets, and aims to help in generating biological insights for specific environments.

Ray is maintained by Sébastien Boisvert, a PhD student supervised by Jacques Corbeil and François Laviolette at Université Laval, in Québec, Canada.

(b) SOAP

SOAP [131–134] has been in evolution from a single alignment tool (available as SOAP v1) to a tool package that provides a complete solution for next generation sequencing data analysis. Currently, SOAP consists of the packages [133] shown in Table 4.4.

Genome sequence assembly tasks were chosen to because of the following reasons:

- **Large problem file size**: Genome Sequencing deals in large data files that is predicted to continue to increase in the future [41]. To sufficiently test whether the hybrid cloud environment can be viable at larger scale computations, it is best to simulate applications that are known to use large data sets.
- **Applications support parallelism**: Both genome sequence assemblers used as a basis for the task simulation (Ray and SOAP) support parallel computations which makes splitting tasks between the multiple resources on the hybrid

Table 4.4 Packages of SOAP

Packages	Descriptions
SOAPaligner/SOAP2	An oligonucleotide alignment tool An efficient program for alignment of short oligonucleotide onto reference sequences It is compatible with single-read or pair-end resequencing
SOAPsnp	A consensus sequence builder It is based on soap1 and SOAPaligner/SOAP2's alignment output It calculates a quality score for each consensus base which can be used for any latter process to call SNPs
SOAPindel	An indel finder It is developed to find the insertion and deletion for resequence technology
SOAPsv	A structural variation scanner It is developed for detecting the structural variation
SOAPdenovo	A tool for short read de novo assembly It is for assembling short oligonucleotide into contigs and scaffolds
SOAP3/GPU	A GPU-accelerated alignment tool It is a GPU-based software for aligning short reads with a reference sequence It can find all alignments with k mismatches, where k is chosen from 0 to 3 When compared with its previous version SOAP2, SOAP3 can be up to tens of times faster

environment much simpler. Features such as error correction and task scheduling are already inbuilt into this application. Other applications may require further development to ensure compatibility with heterogeneous environments such as the hybrid cloud environment being simulated.

4.3   Summary, Discussion and Future Directions

For the future, it is important to focus research on finding ways to further improve performance, efficiency, versatility and automation of the hybrid cloud environment. There are still many other additions and changes to the framework that can be considered to increase its functionality.

4.3.1   What Might the Model Look like?

End users of the environment would get the same interface and usability as a standard Microsoft Azure Cloud environment. The setup should be nearly fully automated. The actual implementation of combining multiple disparate computing resources to form the hybrid cloud environment being used in the back end should, for all intents and purposes, be invisible to the end user.

Figure 4.8 shows a hybrid cloud environment that contains 2 private clouds and a public cloud. This cloud will eventually be expanded to include more available infrastructure. The hybrid cloud environment will need to handle additional

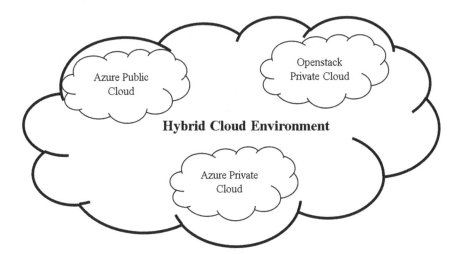

Fig. 4.8 Basic infrastructure of hybrid cloud environment

complexity and effectively manage resources in the system. Current computing systems can provide up to hundreds of thousands of nodes with a small number of homogeneous computational cores per node.

If the hybrid cloud environment is to be scaled to handle exascale problems, the complexity of the computational resources is expected to increase in two dimensions. First, the core count will increase substantially. Second, the cores would be expected to be heterogeneous due to the hybrid nature of the proposed environment. In addition to increasing the complexity of the computational resources, the resources shared between the computational resources can also have a great impact on performance. Therefore a suitable operating system must be selected. Though task management can be handled by the HPC + Cloud task manager, further software considerations that would need to be covered are in the area of fault tolerance and power management. Both factors are expected to become much bigger on the larger scale required.

Increased automation is also expected when setting up the hybrid cloud environment. When the final architecture has been decided and a working case study has been run on the hybrid cloud environment, work can begin on writing installation scripts for the components of the hybrid cloud environment and the control interface GUI.

4.3.2 Application Primitives—Key to Performance

The hybrid cloud environment is expected to include all available computing resources that will be harnessed with software that can divide up the work well. One way to boost performance of the nodes under the hybrid infrastructure is to take advantage of application primitives that will take out the software level on the hardware.

This can be achieved by writing applications that bypass the software layer and directly handles tasks in the firmware layer dependent on the actual hardware of the node that the application is running on. A more generic way to approach the problem is to incorporate a translation layer into the task scheduler that will convert any tasks run on it to the application primitive. By removing a layer when running tasks, performance enhancements are expected. Both approaches will need to be analyzed.

Increased automation is also expected when setting up the hybrid cloud environment. When the final architecture has been decided and a working case study has been run on the hybrid cloud environment, work can begin on writing installation scripts for the components of the hybrid cloud environment and the control interface GUI.

Chapter 5
Optimized Online Scheduling Algorithms

However, all algorithms mentioned in Chap. 2 are considered as concurrent processing but not parallel processing and all are suitable to handle non-data intensive applications in cloud environment as all are considered as complex algorithm which consumes relatively high amount of memory, bandwidth and computational power to maintain its data structure. The outcome of maintaining these data structures will cause the time of scheduling tasks unbounded and make loss in profit gains. Undeniably, profit gain by IaaS provider is inversely proportional to time consumed to finish a task. To encounter most of the aspects and issues which are mentioned in Chap. 3, this project propose an online scheduling algorithm is to overcome the various excessive overheads during process while maintaining service performance and comparable least time consuming approach for data intensive task to adapt in the future cloud system.

5.1 Dynamic Task Splitting Allocator (DTSA)

DTSA is a solution to reduce the stress which rely on work done by single service instance and dynamically divide the big dataset into multiple equal scaled size dataset. Scale factor (SF) is the number of available child level service instances (CLSI) and the master service instance (MSI) is to in charge of splitting the big dataset and equally assign to each of CLSI as shown in Fig. 5.1.

© Springer International Publishing AG 2018   65
R. K. J. Tan et al., *Optimized Cloud Based Scheduling*, Studies in Computational
Intelligence 759, https://doi.org/10.1007/978-3-319-73214-5_5

Fig. 5.1 DTSA model

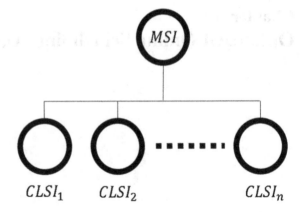

MSI

$CLSI_1$ $CLSI_2$ $CLSI_n$

The details are shown in Algorithm 1.

Algorithm 1: Dynamic Task Splitting Allocator (DTSA)

Input : D : Big Dataset
SF : the number of available CLSI
1. Create equal scaled size datasets $d_i \in$ D , $i > 0$, max (i) = SF
2. Calculate total number of line L in D
3. Declare counter C for current index of located line
4. **for** $\forall$ $d_i \in$ D **do**
5.   **if** $i \neq SF$ **then**
6.   **while** $C < \frac{(L \times i)}{SF}$ **do**
7.   Transfer data of $l_c \in$ L to d_i
8. $C = C + 1$
9.   **end while**
10.   **else**
11.   Transfer the rest data to d_i
12.   **end if**
13. **end for**
14. Assign *each $d_i \in$ D to each $s_i \in$ S*

DTSA algorithm is activated or reset when either each of the following events is triggered:

- A Big Dataset is received for performing tasks
- Any interrupt on Big Dataset on changes from either IaaS service provider or users.

Dynamic and flexible algorithm is to provide an adaptive environment to face the challenge of random events which is important to minimize loss and produce an automated load-balancing mechanism.

DTSA checks all equal numbers of scaled dataset (the outer for loop, Step 4 to 13), if the current index of dataset is not the last index (Step 5) then the counter C will be calculated to determine whether the current part of main dataset belongs to current dataset (Step 6). If it is correct, the data within the line will be transferred from main dataset to current dataset, otherwise next dataset will be in turn (Step 7–9). The rest data belongs to last dataset (Step 11). DTSA will assign each scaled dataset to each of the service instances S respectively (Step 14).

5.2 Procedural Parallel Scheduling Heuristic (PPSH)

Individual computational node with multi-core processors built-in has become massively popular and more sophisticated in this technology-boosted era.

As the results, sequential task scheduling has a typically declined trend of performance in terms of parallelism limits and physical resource utilization. Parallel task scheduling on multi-core node provides a promising method to utilize physical resources in a relatively high comprehensive ratio for producing optimal processing times to maintain profit gains and others when possible.

This project thereby proposes PPSH to process data intensive applications in multi-core processing environment and each service unit is addressed as core level service instance (CLSI). The node level service instance (NLSI) is the MSI of CLSI which is in charge of managing the sequence of procedure and store information and datasets.

PPSH has total four procedures to process big dataset after DTSA assigning each scaled dataset to each CLSI and all procedures process in chronological order which re-utilize a set of available CLSI in parallel repeatedly as shown in Fig. 5.2.

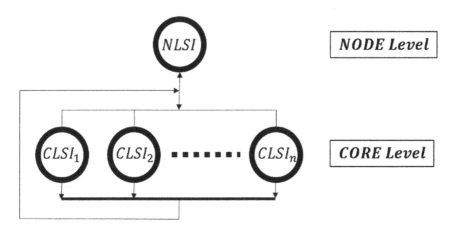

Fig. 5.2 PPSH workflow

Unlike [38], all of scaled datasets are located at NLSI and each of its CLSI is only required to locally access its scaled dataset once and map the required data in own local memory for further processing from PPSH Phase 1 to Phase 4 without accessing the scaled dataset in iteration to reduce communication overheads.

PPSH involves the following four procedures:

5.2.1 Mapping Phase

Extract all the useful key data pairs from every row of scaled datasets and map into memory for further processing. The actual workouts are defined as the following:

- A key is to represent identity of data field

$$k \tag{5.1}$$

- A data is to represent the information of key

$$dt \tag{5.2}$$

- A key data pair (K) is to represent the data is assigned to the key.

$$K = \{k : dt\} \tag{5.3}$$

- A set of key data pair, key data pair collection (KD) is to represent each row of each scaled dataset.

$$KD = \{k_1 : dt_1, k_2 : dt_2, \ldots, k_n : dt_n\} \tag{5.4}$$

- A set of key data pair collection, total key data pair (TKD) is to represent total rows of each scaled dataset.

$$TKD = \{KD_1, KD_2, \ldots, KD_n\} \tag{5.5}$$

- A set of total key data pair, overall of total key data pair $(OTKD)$ is to represent all rows from all scaled dataset available.

$$OTKD = \{TKD_1, TKD_2, \ldots, TKD_n\} \tag{5.6}$$

Details are shown in Algorithm 2.

PPSH Phase 1—Mapping Phase receives a collection of scaled datasets D and starts to activate parallel processing execution on all the $CLSI_{1\ to\ N}$ at the same time (Step 1 to all the remaining steps).

If each of the $CLSI_{n \in N}$ is instance that under control by NLSI and ready for the execution (Step 2–11), memory will be allocated to TKD_n only for $CLSI_n$ Step 3).

There is a checking to confirm that current file pointer is not end of file on scaled dataset $d_n \in D$ then the process will be continue (Step 4–10).

The mapping process will be determined whether the necessary keys are matched and KDs will be allocated memory for storing each mapped data to each key (Step 5–9). Lastly, the counter of KD increases by one to indicate one KD has been successfully mapped in $CLSI_n$ which is under TKD_n (Step 8).

Algorithm 2: PPSH Phase 1—Mapping Phase

Input: N : Number of CLSI available
$Size_{TKD}$: Memory size to store TKD
$Size_{KD}$: Memory size to store KD
C_{TKD_n} : Counter of KD in TKD_n
D : Collection of scaled datasets

1. Activate parallel processing execution on $CLSI_{1\ to\ N}$
2. **if** $CLSI_1 \in NLSI , CLSI_1 \neq \emptyset$ **then**
3.   Allocate memory for $Mem_{TKD_1} \in Size_{TKD}$
4.   **while not** $eof_{d_1 \in D}$ **do**
5.   **if** $\forall k\ are\ matched$ **then**
6.   Allocate memory for $Mem_{KD} \in Size_{KD}$
7.   Map $KD\ according\ to\ each\ k$
8. $C_{TKD_1} = C_{TKD_1} + 1$
9.   **end if**
10.   **end while**
11. **end if**

          ⋮

    if $CLSI_N \in NLSI , CLSI_N \neq \emptyset$ **then**
         Allocate memory for $Mem_{TKD_N} \in Size_{TKD}$
         while not $eof_{d_N \in D}$ **do**
              if $\forall k\ are\ matched$ **then**
                   Allocate memory for $Mem_{KD} \in Size_{KD}$
                   Map $KD\ according\ to\ each\ k$
                   $C_{TKD_N} = C_{TKD_N} + 1$
              end if
         end while
    end if

5.2.2 Shuffling Phase

Rearrange and combine all the KDs in each of TKD_n which KDs are grouped by primary key to further reduce and concentrate the scope of data intensive task running on $CLSI_n$. This can be explained as the following:

- A primary key is a most unique key in KD.

$$PK = \{pk : dt\} \tag{5.7}$$

- It has an irreplaceable attribute during task processing to represent the entire KD such as date, student ID, phone number and others.
- Based on set theory, PK is involved in KD because one of the keys in KD is primary key.

$$PK \subset KD \tag{5.8}$$

$$\{pk : dt\} \subset \{k_1 : dt_1, k_2 : dt_2, \ldots, k_n : dt_n\} \tag{5.9}$$

- Primary key of different KDs can be identical or non-identical.
- All KDs with identical primary key will be grouped together and non-primary key fields will be allocated in a larger memory space.
- The purpose of shuffling phase is to reduce the quantity of KDs so that data intensive process has a more concentrated task which focuses on combined KDs with primary key but the memory size of each combined KD will be increased.
- A set of key data pair, key data pair collection with primary key after combination will result

$$
\begin{aligned}
KD_{ShufflingPhase} = \{ & pk : dt, \\
& k_1 : dt_{1,1}, \ldots, dt_{1,n}, \\
& k_2 : dt_{2,1}, \ldots, dt_{2,n}, \\
& \ldots, \\
& k_n : dt_{n,1}, \ldots, dt_{n,n} \}
\end{aligned}
\tag{5.10}
$$

$KD_{ShufflingPhase}$ is to represent each combined row of each scaled dataset.
- A set of key data pair collection, total key data pair

$TKD_{ShufflingPhase} = \{KD_1, KD_2, \ldots, KD_n\}$ has a reduced amount of KDs which can be represented

$$TKD_{ShufflingPhase} < TKD_{MappingPhase} \tag{5.11}$$

Details are shown in Algorithm 3. PPSH Phase 2—Shuffling Phase re-utilize $CLSI_{1\ to\ N}$ again to activate parallel processing execution (Step 1 to the remaining steps). For each $KD \in Mem_{TKD_n}$ which retrieved from the PPSH Phase 1, KD with first entry of primary key will be allocated memory to store all the current data and subsequent KD with identical primary key (Step 6–16). KD with different primary key will be recognized as new entry and counter will be increased by 1 (Step 13–14).

Algorithm 3: PPSH Phase 2—Shuffling Phase

Input: N : Number of CLSI available

$Size_{TKD}$: Memory size to store TKD which contains combined KDs

$Size_{KD}$: Memory size to store each combined KD

$C^*_{TKD_n}$: Counter of combined KD in TKD_n for Shuffling Phase

$Mem_{TKD_{1\ to\ N}}$: Memory collection of TKD for $\forall CLSI_{1\ to\ N}$
                       recorded in Mapping Phase

1. Activate parallel processing execution on $CLSI_{1\ to\ N}$
2. **if** $CLSI_1 \in NLSI$, $CLSI_1 \neq \emptyset$ **then**
3.   Allocate memory for $Mem^*_{TKD_1} \in Size_{TKD}$
4.   **for** *each* $KD \in Mem_{TKD_1}$ **do**
5.   Next Primary Key:
6.   **if** *first entry of primary key* **then**
7.   Allocate memory for $Mem^*_{KD} \in Size_{KD}$
8.   Shuffle KD *according to primary key*
9.   **else**
10.   **if** *primary key is identical* **then**
11.   Shuffle KD *according to primary key*
12.   **else**
13. $C^*_{TKD_1} = C^*_{TKD_1} + 1$
14.   **goto** Next Primary Key
15.   **end if**
16.   **end if**
17.   **end for**
18. **end if**

                     ⋮

if $CLSI_N \in NLSI$, $CLSI_N \neq \emptyset$ **then**
   Allocate memory for $Mem^*_{TKD_N} \in Size_{TKD}$
   for *each* $KD \in Mem_{TKD_N}$ **do**
     Next Primary Key:
     if *first entry of primary key* **then**
       Allocate memory for $Mem^*_{KD} \in Size_{KD}$
       Shuffle KD *according to primary key*
     else
       if *primary key is identical* **then**
         Shuffle KD *according to primary key*
       else
         $C^*_{TKD_N} = C^*_{TKD_N} + 1$
         goto Next Primary Key
       end if
     end if
   end for
end if

Appendix: Simulation Results for Sect. 5.2
A.1 Single Node
See Fig. A.1.

Comparisons Between MultiCore and MultiNode Processing on Single Node							
Input Data Size	10 GB						
Processor Brand of Node	AMD Opteron Processor 4171 HE 2.10 GHz (2 Physical Cores , 8 Logical Cores)						
Total Number of Physical Nodes in Parallel	1	1	1	1	1	1	1	1
Total Number of Logical Cores in Parallel	1	2	3	4	5	6	7	8
Total RAM (GB)	56	56	56	56	56	56	56	56
LAN Connection Between Nodes	10 Gbps						
Splitting Process Time (s)	918.2	981.25	915.93	916.61	956.02	909.52	1652.87	1092.76
Mapping Process Time (s)	1596.88	1351.39	959.66	727.02	618.3	523.7	457.75	416.44
Shuffling Process Time (s)	987.47	1134.17	840.98	882.7	1068.9	1107.61	1193.33	1895
Reducing Process Time (s)	72.72	66.98	159.23	163.53	49.34	110.31	131.35	117.62
Execution Time (s)	3575.51	3533.97	2877.86	2690.53	2693.11	2651.45	3435.84	3526.4

Fig. A.1 Simulation results of single node with multi-core

A.2 Two Nodes
See Fig. A.2.

Comparisons Between MultiCore and MultiNode Processing on Two Nodes							
Input Data Size	10 GB						
Processor Brand of Node	AMD Opteron Processor 4171 HE 2.10 GHz (2 Physical Cores, 8 Logical Cores)						
Total Number of Physical Nodes in Parallel	2	2	2	2	2	2	2	2
Total Number of Logical Cores in Parallel	2	4	6	8	10	12	14	16
Total RAM (GB)	56 X 2	56 X 2	56 X 2	56 X 2	56 X 2	56 X 2	56 X 2	56 X 2
LAN Connection Between Nodes	10 Gbps						
Splitting Process Time (s)	984.93	970.52	964.31	983.26	969.79	898.58	982.71	966.81
Mapping Process Time (s)	807.51	686.6	481.51	360.8	308.74	261.51	229.54	206.11
Shuffling Process Time (s)	62.5	62.95	78.79	115.55	287.22	361.32	434.67	560.19
Reducing Process Time (s)	0.69	0.28	0.45	0.62	0.59	0.63	0.7	0.61
Execution Time (s)	1857.82	1723.26	1533.16	1468.02	1572.85	1527.64	1651.74	1820.34

Fig. A.2 Simulation results of two nodes with multi-core

A.3 Three Nodes
See Fig. A.3.

Comparisons Between MultiCore and MultiNode Processing on Three Nodes							
Input Data Size	10 GB						
Processor Brand of Node	AMD Opteron Processor 4171 HE 2.10 GHz (2 Physical Cores, 8 Logical Cores)						
Total Number of Physical Nodes in Parallel	3	3	3	3	3	3	3	3
Total Number of Logical Cores in Parallel	3	6	9	12	15	18	21	24
Total RAM (GB)	56 X 3	56 X 3	56 X 3	56 X 3	56 X 3	56 X 3	56 X 3	56 X 3
LAN Connection Between Nodes	10 Gbps						
Splitting Process Time (s)	983.29	971.36	983.73	970.29	979.54	943.97	945.36	939.78
Mapping Process Time (s)	536.1	461.94	317.75	244.09	199.89	169.9	152.18	135.79
Shuffling Process Time (s)	63.21	66.61	72.26	90.33	149.87	294.93	476.96	683.1
Reducing Process Time (s)	0.38	0.2	0.17	0.27	0.88	0.58	0.44	0.51
Execution Time (s)	1585.48	1505.54	1377.68	1313.6	1337.61	1416.45	1585.64	1763.62

Fig. A.3 Simulation results of three nodes with multi-core

5.2.3 Eliminating Phase

A KD has multiple keys and data but there is no requirement to utilize most of it. Keeping most of the redundant data will tremendously affect the performance of IaaS cloud. As the result, creating an algorithm to filter out unnecessary keys and data by following the filtering factors given. This can be explained as the following:

- Filtering Factor can be single or multiple depending on the instruction received.

$$FF = \{f_1, f_2, \ldots, f_n\} \tag{5.12}$$

- The keys and data will be filtered out by every single factor in the set of FF and the most valuable keys and data will be recorded.
- A set of key data pair, key data pair collection with primary key after elimination process will result the

$$
\begin{aligned}
KD_{EliminatingPhase} = \{ &pk : dt, \\
&k_1 : dt_{1,1}, \ldots, dt_{1,n}, \\
&k_2 : dt_{2,1}, \ldots, dt_{2,n}, \\
&\ldots, \\
&k_n : dt_{n,1}, \ldots, dt_{n,n} \}
\end{aligned} \tag{5.13}
$$

where

$$KD_{ElimitatingPhase} < KD_{ShufflingPhase} \tag{5.14}$$

and

$$TKD_{EliminatingPhase} < TKD_{ShufflingPhase} \tag{5.15}$$

- The effect of elimination is based on the strength of the FF, the more stringent FF is, the n number of key data pair will be reduced effectively. Details are shown in Algorithm 4.

PPSH Phase 3—Eliminating Phase continuously re-utilize the $CLSI_{1 \text{ to } N}$ to activate parallel processing execution (Step 1–11). The filtering factors play a major role to decide whether the KD is suitable according to primary key and KD recording will be continuously increased once it is matched to the FF (Step 5–8).

Algorithm 4: PPSH Phase 3—Eliminating Phase

Input: N : Number of CLSI available

$Size_{TKD}$: Memory size to store TKD which contains the
total remaining KDs after elimination

$Size_{KD}$: Memory size to store each KD after elimination

$Mem_{TKD_{1\,to\,N}}$: Memory collection of TKD for $\forall CLSI_{1\,to\,N}$
recorded in Shuffling Phase

$C^*_{TKD_n}$: Counter of KD in TKD_n for Eliminating Phase

FF : A set of Filtering Factor

1. Activate parallel processing execution on $CLSI_{1\,to\,N}$

2. **if** $CLSI_1 \in NLSI$, $CLSI_1 \neq \emptyset$ **then**
3. | Allocate memory for $Mem^*_{TKD_1} \in Size_{TKD}$

4. | **for** *each* $KD \in Mem_{TKD_1}$ **do**

5. | | **if** *KD matched FF* **then**
6. | | | Allocate memory for $Mem^*_{KD} \in Size_{KD}$
7. | | | Record *KD according to Primary Key*
8. | | | $C^*_{TKD_1} = C^*_{TKD_1} + 1$
9. | | **end if**

10. | **end for**

11. **end if**

$\vdots$

if $CLSI_N \in NLSI$, $CLSI_N \neq \emptyset$ **then**
| Allocate memory for $Mem^*_{TKD_N} \in Size_{TKD}$

| **for** *each* $KD \in Mem_{TKD_N}$ **do**

| | **if** *KD matched FF* **then**
| | | Allocate memory for $Mem^*_{KD} \in Size_{KD}$
| | | Record *KD according to Primary Key*
| | | $C^*_{TKD_N} = C^*_{TKD_N} + 1$
| | **end if**

| **end for**

end if

5.2.4 Finishing Phase

This phase can be considered as further elimination stage to finish the task and the optimal final information will be revealed. NLSI as well as the MSI of $CLSI_{1\ to\ N}$ in charge of collecting the $TKD_{1\ to\ N}$ and analyse it by FF again. Details are shown in Algorithm 5.

PPSH Phase 4—Finishing Phase verify whether the NLSI is their MSI and allow it to collect TKD from each of CLSI (Step 1–2). To analyse key data pair thoroughly, every KD from every TKD will be checked whether it is matched the FF set by NLSI (Step 3–9). This phase is based on sequential programming to check every KD in workflow, as the majority of data has already been eliminated by 3 Phases above, the remaining quantity of data in this phase will not affect the consumed time in large fluctuation.

Algorithm 5: PPSH Phase 4—Finishing Phase

Input: N : Number of CLSI available
$Mem_{TKD_{1\ to\ N}}$: Memory collection of TKD for $\forall CLSI_{1\ to\ N}$
                recorded in Eliminating Phase
FF : A set of Filtering Factor

1. NLSI collects Mem_{OTKD} from all individual Mem_{TKD_n}
2. **if** $NLSI \in pred(CLSI_{1\ to\ N})$, $NLSI \neq \emptyset$ **then**
3.   **for** each $Mem_{TKD_n} \in Mem_{OTKD}$ **do**
4.   **for** each $KD \in Mem_{TKD_n}$ **do**
5.   **if** KD matched FF **then**
6.   Output the result of key data pair
7.   **end if**
8.   **end for**
9.   **end for**
10. **end if**

5.3 Scalable Parallel Scheduling Heuristic (SPSH)

Large scale cloud computing is emerging and this includes geographically distributed data centres with heterogeneous cloud model from different cloud service providers. The proposed SPSH is mainly targeting on high scalability concept with tree-based design for ease of task monitoring, result collection and scheduling management from MSI to process large-scale parallel and optimal performance computing tasks in every sublevel. IaaS cloud can be expanded and classified into four levels which refer to Fig. 5.3.

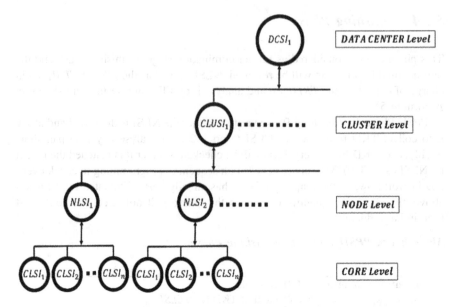

Fig. 5.3 Scalable levels in IaaS cloud environment

- Core Level (CL): The base level of cloud which is derived from multi-core processor as basic service unit. Each processor core can be a service instance which is called Core Level Service Instance (CLSI),

$$\mathrm{CL} = \{CLSI_1, CLSI_2, \ldots, CLSI_n\} \tag{5.16}$$

- Node Level (NL): The second level of cloud which is derived from node as basic service unit. Each node contains one or more than one multi-core processors. Each node is called Node Level Service Instance (NLSI),

$$NL = \{NLSI_1, NLSI_2, \ldots, NLSI_n\} \tag{5.17}$$

$$CL \subset NL \tag{5.18}$$

- Cluster Level (CLUL): The third level of cloud which is derived from a set of nodes as basic service unit. Each cluster is called Cluster Level Service Instance (CLUSI),

$$CLUL = \{CLUSI_1, CLUSI_2 \ldots, CLUSI_n\} \tag{5.19}$$

$$NL \subset CLUL \tag{5.20}$$

- Data Centre Level (DCL): The top level of cloud which is derived from a set of clusters as basic service unit. Each data centre is called Data Centre Service Instance (DCSI),

$$DCL = \{DCSI_1, DCSI_2, \ldots, DCSI_n\} \tag{5.21}$$

$$CLUL \subset DCL \tag{5.22}$$

SPSH is an algorithm which is based on the four levels described above and the whole structure can be analysed with precedence-constraints as the following:

$$CL \subset NL \subset CLUL \subset DCL \tag{5.23}$$

Details are shown in Algorithm 6.

Algorithm 6 : Scalable Parallel Scheduling Heuristic (SPSH)

Input : N_{DCL} : Number of Data Centres
$D_{Original}$: Original Big Dataset From Service User

1. Execute DTSA with N_{DCL}
2. Parallel execution starts at DCL

3. **if** $DCSI_1 \in DCL, DCSI_1 \neq \emptyset$ **then**
4.   Execute DTSA with N_{CLUL}
5.   Parallel execution starts at CLUL

6.   **if** $CLUSI_1 \in CLUL, CLUSI_1 \neq \emptyset$ **then**
7.   Execute DTSA with N_{NL}
8.   Parallel execution starts at NL

9.   **if** $NLSI_1 \in NL, NLSI_1 \neq \emptyset$ **then**
10.   Execute DTSA with N_{CL}
11.   Execute PPSH
12.   **end if**
              ⋮

a.   **if** $NLSI_n \in NL, NLSI_n \neq \emptyset$ **then**
b.   Execute DTSA with N_{CL}
c.   Execute PPSH
d.   **end if**

e.   Select most optimal set of result from $NLSI_{1\,to\,n}$
f.   **end if**
          ⋮

g.   **if** $CLUSI_n \in CLUL, CLUSI_n \neq \emptyset$ **then**
          ⋮

h.   **end if**

i.   Select most optimal set of result from $CLUSI_{1\,to\,n}$
j. **end if**

k. **if** $DCSI_n \in DCL, DCSI_n \neq \emptyset$ **then**
      ⋮

l. **end if**

m. Select most optimal set of result from $DCSI_{1\,to\,n}$
n. Output the set of result to service user

SPSH is a solution of dynamic programming that repeatedly breaks down the $D_{Original}$ into N numbers with each size S

$$N = N_{DCL} \times N_{CLUL} \times N_{NL} \times N_{CL} \qquad (5.24)$$

$$S = \frac{D_{Original}}{N_{DCL} \times N_{CLUL} \times N_{NL} \times N_{CL}} \qquad (5.25)$$

and execute in parallel level by level until all of the D_S are solved. The next step will be combination results of D_S and the final result of $D_{Original}$ will be revealed completely in every facet.

In Algorithm 6, SPSH executes DTSA to split $D_{Original}$ evenly to each $DCSI_x \in DCL$ (Step 1) and starts parallel execution on all $DCSI_{1ton} \in DCL$ (Step 3 to Step j for $DCSI_1$, Step k to l for $DCSI_n$). Compare each of $DCSI_{1\ to\ n}$ and select the most optimal result out of it and output to service user (Step m–n).

In case of $DCSI_1$, it executes DTSA to split $\frac{D_{Original}}{N_{DCL}}$ evenly to each $CLUSI_x \in CLUL$ (Step 4) and starts parallel execution on all $CLUSI_{1\ to\ n} \in CLUL$ (Step 6 to Step f for $CLUSI_1$, Step g to Step h for $CLUSI_n$. Select the most optimal set of result from $CLUSI_{1\ to\ n}$ (Step i).

In case of $CLUSI_1$, it executes DTSA to split $\frac{D_{Original}}{N_{DCL} \times N_{CLUL}}$ evenly to each $NLSI_x \in NL$ (Step 7) and starts parallel execution on all $NLSI_{1\ to\ n} \in NL$ (Step 9–12 for $NLSI_1$, Step a to Step d for $NLSI_n$. Select the most optimal set of result from $NLSI_{1\ to\ n}$ (Step e).

In case of $NLSI_1$, it executes DTSA to split $\frac{D_{Original}}{N_{DCL} \times N_{CLUL} \times N_{NL}}$ evenly (Step 10) and executes PPSH (Step 11).

5.4 Calculation of Time

There are two types of process time calculation in the proposed algorithm which are splitting time on MSI (ST_{MSI}) and execution time on $CLSI_{1ton}$ ($ET_{CLSI_{1\ to\ n}}$). The overheads include communication time between service instances X and Y ($C_{X,Y}$) and service instance passive delay on service instance X ($SIPD_X$). SIPD is passive because all the service instances located in same level must be started and end at the same time to adapt parallelism concept. If one of the service instance delay the end time due to some reasons, the rest will have to delay as well. SIPD has minimal impact to affected end time if all the service instances have same computation performance and same size of received dataset. Figure 5.4 has shown in details.

ST_{MSI} can be categorized as the following:

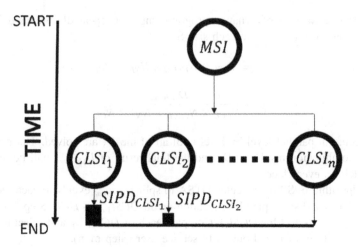

Fig. 5.4 Service instance passive delay (SIPD)

- **Earliest Start Splitting Time ($ESST_{MSI}$)**

$$ESST_{MSI} = 0 \qquad (5.26)$$

For initialization purpose.
- **Earliest Finish Splitting Time ($EFST_{MSI}$)**

$$EFST_{MSI} = \min_{X \in MSI} \left(ESST_X + ST_X \right) \qquad (5.27)$$

- **Latest Start Splitting Time ($LSST_{MSI}$)**

$$LSST_{MSI} = \max_{Y \in pred(MSI)} \left(LFET_Y + C_{Y,MSI} \right) \qquad (5.28)$$

- **Latest Finish Splitting Time ($LFST_{MSI}$)**

$$LFST_{MSI} = \max_{X \in MSI} \left(LSST_X + ST_X \right) \qquad (5.29)$$

$ET_{CLSI_{1ton}}$ can be categorized as the following:
- **Earliest Start Execution Time ($ESET_{CLSI_{1\ to\ n}}$)**

$$ESET_{CLSI_{1\ to\ n}} = \min_{X \in MSI, Z \in each(CLSI_{1\ to\ n})} \left(EFST_X + C_{X,Z} \right) \qquad (5.30)$$

- **Earliest Finish Execution Time ($EFET_{CLSI_{1\ to\ n}}$)**

$$EFET_{CLSI_{1 \, to \, n}} = \min_{Z \in each(CLSI_{1 \, to \, n})} (ESET_Z + ET_Z) \qquad (5.31)$$

- **Latest Start Execution Time ($LSET_{CLSI_{1 \, to \, n}}$)**

$$LSET_{CLSI_{1 \, to \, n}} = \max_{X \in MSI, Z \in each(CLSI_{1 \, to \, n})} (LFST_X + C_{X,Z}) \qquad (5.32)$$

- **Latest Finish Execution Time ($LFET_{CLSI_{1 \, to \, n}}$)**

$$LFET_{CLSI_{1 \, to \, n}} = \max_{Z \in each(CLSI_{1 \, to \, n})} (LSET_Z + ET_Z + SIPD_Z) \qquad (5.33)$$

Chapter 6
Performance Evaluation

This chapter presents the evaluation of the proposed algorithms to verify the effectiveness. Thus, first analyse the time complexity of or proposed algorithms and present the experimental results.

6.1 Complexity Analysis

6.1.1 Algorithm 1: Dynamic Task Splitting Allocator (DTSA)

Time complexity from Step 6 to 9 is $O(n)$, from Step 4 to 13 is $O(iC)$ for i numbers of dataset to be allocated to same number of service instances equally and C numbers of counter. Defining that $O(iC) < O(n^2)$ because the number of scaled datasets which will be assigned to same number of service instances. The quantity of service instances on IaaS cloud is always limited and predictable.

6.1.2 Algorithm 2: PPSH Phase 1—Mapping Phase

Time complexity of this algorithm is $N \times O(n)$ as N numbers of service instance process in parallel at the same time to map scaled datasets.

© Springer International Publishing AG 2018
R. K. J. Tan et al., *Optimized Cloud Based Scheduling*, Studies in Computational Intelligence 759, https://doi.org/10.1007/978-3-319-73214-5_6

6.1.3   Algorithm 3: PPSH Phase 2—Shuffling Phase

From Step 4 to 17, it is to determine time complexity of this algorithm on single service instances $O(n)$ as well as N numbers of service instance $N \times O(n)$ running in parallel to shuffle the KD according to its primary key.

6.1.4   Algorithm 4: PPSH Phase 3—Eliminating Phase

Eliminating process focuses on each KD which is matched the FF and record accordingly. Time complexity of this algorithm is $N \times O(n)$ as N numbers of service instance process in parallel at the same time.

6.1.5   Algorithm 5: PPSH Phase 4—Finishing Phase

NLSI collect the TKDs from N number of service instances and check X number of KDs from every collected TKD under FF conditions so that the time complexity is $O(NX)$ where $O(NX) < O(n^2)$ as N number of service instances is countable in advance.

6.1.6   Algorithm 6: Scalable Parallel Scheduling Heuristic (SPSH)

The proposed SPSH is overhead-free tree-based design because there is no requirement to spend additional resources on maintaining the tree structure as introduced in Algorithm 6 so that the size of big dataset given will not increase the additional unnecessary overhead generated. From the analysis above, there is a conclusion that the time complexity of DTSA is $O(iC)$ and overall PPSH is sum of $N \times O(n)$, $N \times O(n)$, $N \times O(n)$ and $O(NX)$ which can be simplified as $O(NX)$. For each DCSI, it is needed to analyse the time complexity from NL.

For each $NLSI \in NL$, time complexity is $O(iC) + O(NX)$ from Step 9 to 12 and N numbers of NLSI in parallel can be simplified as same as $O(iC) + O(NX)$ from Step 9 to Step d.

For each $CLUSI \in CLUL$, time complexity is $O(iC) + O(iC) + O(NX)$ which can be simplified as $O(iC) + O(NX)$ as well from Step 6 to Step f. N numbers of CLUSI can be simplified to $O(iC) + O(NX)$ from Step 6 to Step h.

Therefore, for each of $DCSI \in DCL$ the time complexity is $O(iC) + O(iC) + O(NX)$. After simplification and consideration of DCSI with N numbers of parallel

Table 6.1 Comparisons between SPSH and algorithms on data intensive task

Algorithm	SPSH	Profit-based scheduling [109]	Preemptable scheduling [110]	Priority-based scheduling [111]
Main strength	+High scalability +Dynamic – Load-balancing	+Dynamic –Load-balancing +Profit maximization	+Preemption for higher priority task +Profit maximization	+Profit maximization +High priority for critical path
Main weakness	–High ownership cost –Non-preemptable execution	–Some service instances in idle state during execution –High communication overheads	–High overheads during pre-emption –Complicated task execution	–Difficult to maintain the structure –Huge overheads
Most optimal case	$O(iC) + O(NX)$	$O(n^2)$	$O(n^2)$	$O(n^2)$
Average case	$O(iC) + O(NX)$	$O(n^2)$	$O(n^2)$	$O(n^2)$
Worst case	$O(iC) + O(NX)$	$O(n^2)$	$O(n^2)$	$O(n^2)$

quantity the final overall time complexity from Step 1 to the rest of SPSH is $O(iC) + O(NX)$ which can be represented as the following:

$$O(iC) + O(NX) < O\left(n^2\right) \qquad (6.1)$$

Table 6.1 has shown the comparisons between SPSH and the discussed algorithms in Chap. 2.

6.2 Experimental Results

As the result from complexity analysis, the experimental results which are obtained from SPSH simulation of size 10 GB dataset. There are three node level service instances involved in simulation running on Windows Azure Virtual Machines, all are 8-core servers, with 2 AMD Opteron Processor 4171 HE 2.10 GHz processors, 56 GB memory, 10 Gbps network connection and 300 GB disk space per node.

Figure 6.1 shows the performance of multi-core processing from single core sequential processing to 8-core parallel processing. 6-core parallel processing has outstanding improvement compared to others and 7–8-core processing has drastically declined performance in execution time but still better than single core. Table 6.2 has shown the results of comparisons between N-core and N-core parallel

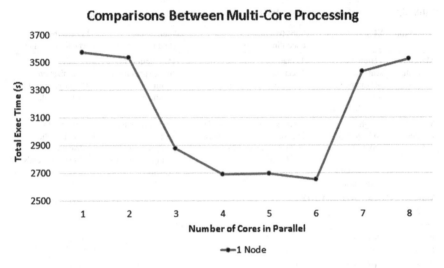

Fig. 6.1 Parallel performance in single node

Table 6.2 Performance comparisons between N-core to N-core processing

Cores	Cores							
	1 (%)	2 (%)	3 (%)	4 (%)	5 (%)	6 (%)	7 (%)	8 (%)
1	0.00	−1.18	−24.24	−32.89	−32.77	**−34.85**	−4.07	−1.39
2	1.16	0.00	−22.80	−31.35	−31.22	**−33.28**	−2.86	−0.21
3	19.51	18.57	0.00	−6.96	−6.86	**−8.54**	16.24	18.39
4	24.75	23.87	6.51	0.00	0.10	**−1.47**	21.69	23.70
5	24.68	23.79	6.42	−0.10	0.00	**−1.57**	21.62	23.63
6	**25.84**	**24.97**	**7.87**	**1.45**	**1.55**	**0.00**	**22.83**	**24.81**
7	3.91	2.78	−19.39	−27.70	−27.58	**−29.58**	0.00	2.57
8	1.37	0.21	−22.54	−31.07	−30.94	**−33.00**	−2.64	0.00

processing and 6-core processing result has all value positive which is a proven record.

Figure 6.2 has shown the performance comparisons between single node and two nodes on multi-core parallel processing. The execution time has a nearly half time reduction in every single core processing in two nodes due to the double increase of RAM and the most optimal performance falls to 4-core instead of 6-core because of the rising demand of network communication control. Table 6.3 has shown the performance comparisons between single node and two nodes parallel execution with multi-core involved.

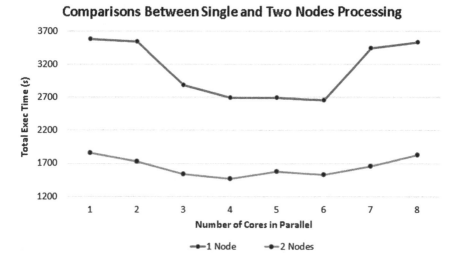

Fig. 6.2 Parallel performance between single and two nodes

Table 6.3 Performance comparisons between single node N-core to two nodes N-core processing

Two nodes processing—cores	Single node processing—cores							
	1 (%)	2 (%)	3 (%)	4 (%)	5 (%)	6 (%)	7 (%)	8 (%)
1	48.04	47.43	35.44	30.95	31.02	29.93	45.93	47.32
2	51.80	51.24	40.12	35.95	36.01	35.01	49.84	51.13
3	57.12	56.62	46.73	43.02	43.07	42.18	55.38	56.52
4	**58.94**	**58.46**	**48.99**	**45.44**	**45.49**	**44.63**	**57.27**	**58.37**
5	56.01	55.49	45.35	41.54	41.60	40.68	54.22	55.40
6	57.27	56.77	46.92	43.22	43.28	42.38	55.54	56.68
7	53.80	53.26	42.61	38.61	38.67	37.70	51.93	53.16
8	49.09	48.49	36.75	32.34	32.41	31.35	47.02	48.38

Figure 6.3 has shown the performance between two nodes and three nodes execution with multi-core involved. The comparison results is not as novel as the result between single and two nodes because the RAM is adequate to fulfil the task execution requirement but there is a similar part which is the most optimal

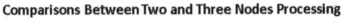

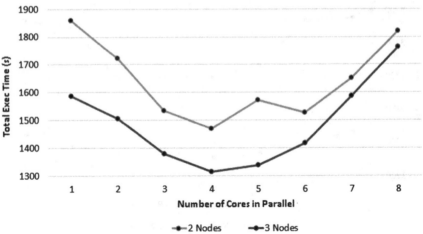

Fig. 6.3 Parallel performance between two and three nodes

Table 6.4 Performance comparisons between two nodes N-core to three nodes N-core processing

Three nodes processing—cores	Two nodes processing—cores							
	1 (%)	2 (%)	3 (%)	4 (%)	5 (%)	6 (%)	7 (%)	8 (%)
1	14.66	8.00	−3.41	−8.00	−0.80	−3.79	4.01	12.90
2	18.96	12.63	1.80	−2.56	4.28	1.45	8.85	17.29
3	25.84	20.05	10.14	6.15	12.41	9.82	16.59	24.32
4	**29.29**	**23.77**	**14.32**	**10.52**	**16.48**	**14.01**	**20.47**	**27.84**
5	28.00	22.38	12.75	8.88	14.96	12.44	19.02	26.52
6	23.76	17.80	7.61	3.51	9.94	7.28	14.24	22.19
7	14.65	7.99	−3.42	−8.01	−0.81	−3.80	4.00	12.89
8	5.07	−2.34	−15.03	−20.14	−12.13	−15.45	−6.77	3.12

performance execution time is on 4-core processing again. It is a significant proof to confirm that 4-core processing has the most optimal results in multi-node and multi-core parallel processing. Table 6.4 has shown the comparison details.

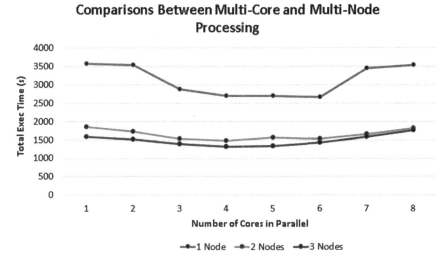

Fig. 6.4 Simulation results of SPSH

In conclusion, as shown in Fig. 6.4, in node level, the more nodes are processing in parallel, the execution time will be reduced accordingly but there is a limit in core level as the shortest execution time is limited to 4-core processing in parallel. It is because the remaining cores are required to run background processes e.g. operating system, network management and others.

Chapter 7
Conclusion and Future Works

Recent technological advancement have emerged the data intensive tasks in wide distinctive areas such as commerce, earth science, astronomy, computational biology and others in boosting trend. Due to the various characteristics of data, it is needed to develop a new data processing architecture for data acquisition, data analysis, data mining, data transmission between service instances, data storage and others, aiming to schedule it into a series of scalable tasks with holistic, viable and modern approach.

Cloud computing is a new generation of computing platform, providing on-demand and pay-per-use services with drastically reduced ownership costs and operating costs dealing on the data intensive tasks. IaaS is the backbone of cloud computing which has the busiest data execution traffics imported from multiple channels. Therefore, optimized online scheduling algorithms play a main role on it.

Before designing an effective algorithms, it is needed to plan properly with investigating various aspects to be noticed and the current issues that may lead to a failure of design and development. From the point of view of IaaS service provider, the provision of cloud services to service users with optimal performance but maintaining the profit margin is the major challenge. Besides, the efficient distribution of current limited cloud resources with economical approach is another problem. Moreover, the adaptability to fulfil the random changes from service users while maintaining the operability of cloud services is crucial key. Lastly, the mapping of logical resources and physical resources in optimal combinations. All must be taken during the requirement analysis stage, design stage, implementation stage, testing stage and further improvement in evolution stage.

This project has proposed an optimal tree-based design task scheduling algorithm to achieve the main load-balancing objective of minimizing time, communication overheads, profit loss and maximizing resource utilization on data intensive tasks. This optimized algorithm contains high scalable service instances and high granularity on datasets matched from core level to data centre level. Batch processing of each level with promising parallel execution on multi-key data pairs to provide sets of result for MSI to eliminate unnecessary sets based on FF and

© Springer International Publishing AG 2018
R. K. J. Tan et al., *Optimized Cloud Based Scheduling*, Studies in Computational Intelligence 759, https://doi.org/10.1007/978-3-319-73214-5_7

produce the most optimal single set of result to service users after evaluation processes.

There is another point to be noticed that is the number of parallel executed service instances in each level can be a large quantity and these large amount of service instances may be or may not be in same geographical location, even if those are in same local area network, the network latency may be or may not be equal whereas some are connected with high speed medium such as fibre optics in server room while others are connected with copper cable in different offices. These issue-specific considerations have been compromised into proposed algorithms and an improved idea will be following up in the future works that is to set up a time-out counter to exclude the service instances with underperformance.

The proposed algorithm has passed two performance evaluation stages which are complexity analysis with other three existing algorithms and simulation of data intensive task scheduling with single node, two nodes and three nodes in parallel multi-core executions. The simulation stage produced an overall makespan comparison results and the execution time had been reduced while the number of nodes running in parallel increased. In case of core level, there is a conclusion that it is not possible to fully utilize all the available core to execute tasks in parallel at the same time. It is because there is other background processes such as the services to maintain the workflow of operating system, daemon of applications, network control services and others. Instead, the number of designated cores to be assigned is four as the 4-core parallel processing has the most optimal result in the above-mentioned simulation. As proven result with evaluation of extensive simulations, this proposed algorithm can be considered as a holistic approach to real-world applications.

However, there are many possible future works to be explored such as pre-emptable parallel batch execution, it may produce a better approach to provide more complete resource utilization but also leads to delays in synchronization and overheads in communication and cache storage which affects overall performance of task scheduling.

Besides, data-agnostic scheduling algorithm is a challenging future works to perform data mining due to consideration of global data coherence from multiple heterogeneous clouds as data will not be in certain format. Instead, it will be a lump sum amount of raw data mining in byte with complex combinatorial optimization problems and another issue will be the integration of platform-agnostic data which means the dataset generated by different devices such as mobile phones, embedded systems and operating systems such as Linux, Android and others all will be necessary future works.

Bibliography

1. K.M. Shegda, W. Andrews, K. Chin, M.R. Gilbert, H. Koehler-Kruener, G. Tay, *Predicts 2014: Content Gets Bigger, Richer and More Personal* (Gartner, 2013)
2. N. Heudecker, M.A. Beyer, D. Laney, M. Cantara, A. White, R. Edjlali, *Predicts 2014: Big Data* (Gartner, 2013)
3. G. Bell, J. Gray, A. Szalay, Petascale computational systems. Computer **39**, 110–112 (2006)
4. J. Monroe, A. Chandrasekaran, A. Dayley, V. Filks, S. Zaffos, J. Unsworth et al., *Predicts 2014: More Storage Capacity and Efficiency, Less Cost—Managing Infinite Data From Every Direction* (Gartner, 2013)
5. S. Baghdassarian, B. Blau, J. Ekholm, S. Shen, *Predicts 2014: Apps, Personal Cloud and Data Analytics Will Drive New Consumer Interactions* (Gartner, 22 Nov 2013)
6. C. Milanesi, M. Escherich, H.J.D.L. Vergne, A. McIntyre, J. Ekholm, A. Gupta, *Predicts 2014: Cognizant Computing—Another Kind of Smart Device* (Gartner, Dec 2013)
7. L. Wang, W. Jie, J. Chen, *Grid Computing: Infrastructure, Service, and Applications* (CRC Press, Boca Raton, 2009)
8. A.J. Hey, K.M. Tolle, D.S.W. Tansley, The fourth paradigm: data-intensive scientific discovery. Proc. IEEE **99**, 1334–1337 (2011)
9. E. Afgan, D. Baker, N. Coraor, H. Goto, I.M. Paul, K.D. Makova et al., Harnessing cloud computing with Galaxy Cloud. Nat. Biotechnol. **29**, 972–974 (2011)
10. S. Sakr, A. Liu, D.M. Batista, M. Alomari, A survey of large scale data management approaches in cloud environments. Commun. Surv. Tutorials IEEE **13**, 311–336 (2011)
11. D.M. Smith, G. Petri, Y.V. Natis, D. Scott, M. Warrilow, J. Heiser et al., *Predicts 2014: Cloud Computing Affects All Aspects of IT* (Gartner, 04 Dec 2013)
12. S. Moore, *Gartner Identifies Top Ten Disruptive Technologies for 2008 to 2012*, vol. 7 (Gartner Inc., Last visited July 2008)
13. S. Tai, Cloud service engineering: a service-oriented perspective on cloud computing, in *Towards a Service-Based Internet* (Springer, 2011), pp. 191–193.
14. Microsoft, *Unified Management for the Cloud OS: System Center 2012 R2* (2013, 11 Mar 2014). Available: http://download.microsoft.com/download/7/7/2/7721670F-DEF0-40D3-9771-43146DED5132/System_Center_2012%20R2_Overview_White_Paper.pdf
15. VMware, *Innovate and Thrive in the Mobile-Cloud Era* (2014)
16. D. Jacobs, Cloud computing (introduction), in *Technology Time Machine Symposium (TTM), 2012 IEEE* (2012), pp. 1–41
17. R. Buyya, C.S. Yeo, S. Venugopal, J. Broberg, I. Brandic, Cloud computing and emerging IT platforms: vision, hype, and reality for delivering computing as the 5th utility. Futur. Gener. Comput. Syst. **25**(6), 599–616 (2009)
18. L. Rodero-Merino, L.M. Vaquero, V. Gil, F. Galán, J. Fontán, R.S. Montero et al., From infrastructure delivery to service management in clouds. Futur. Gener. Comput. Syst. **26**(10), 1226–1240 (2010)

© Springer International Publishing AG 2018
R. K. J. Tan et al., *Optimized Cloud Based Scheduling*, Studies in Computational Intelligence 759, https://doi.org/10.1007/978-3-319-73214-5

19. Z. Shuai, C. Xuebin, Z. Shufen, H. Xiuzhen, The comparison between cloud computing and grid computing, in *2010 International Conference on Computer Application and System Modeling (ICCASM)* (2010), pp. V11-72–V11-75

20. E. Walker, Benchmarking Amazon EC2 for high-performance scientific computing. USENIX login **33**, 18–23 (2008)

21. J. Napper, P. Bientinesi, Can cloud computing reach the top500? in *Presented at the Proceedings of the Combined Workshops on Unconventional High Performance Computing Workshop Plus Memory Access Workshop*, Ischia, Italy, 2009

22. K.R. Jackson, L. Ramakrishnan, K. Muriki, S. Canon, S. Cholia, J. Shalf et al., Performance analysis of high performance computing applications on the amazon web services cloud, in *2010 IEEE Second International Conference on Cloud Computing Technology and Science (CloudCom)* (2010), pp. 159–168

23. R.R. Expósito, G.L. Taboada, S. Ramos, J. Touriño, R. Doallo, Performance analysis of HPC applications in the cloud. Futur. Gener. Comput. Syst. **29**(1), 218–229, (2013)

24. C. Evangelinos, C. Hill, Cloud computing for parallel scientific HPC applications: feasibility of running coupled atmosphere-ocean climate models on Amazon's EC2. ratio **2**, 2–34 (2008)

25. J. Ekanayake, G. Fox, High performance parallel computing with clouds and cloud technologies, in *Cloud Computing*. vol. 34, ed. by D. Avresky, M. Diaz, A. Bode, B. Ciciani, E. Dekel (Springer, Berlin, 2010), pp. 20–38

26. A.I. Avetisyan, R. Campbell, I. Gupta, M.T. Heath, S.Y. Ko, G.R. Ganger et al., Open cirrus: a global cloud computing testbed. Computer **43**, 35–43 (2010)

27. S. Manavi, S. Mohammadalian, N.I. Udzir, A. Abdullah, Hierarchical secure virtualization model for cloud, in *2012 International Conference on Cyber Security, Cyber Warfare and Digital Forensic (CyberSec)* (2012), pp. 219–224

28. J. Manyika, M. Chui, B. Brown, J. Bughin, R. Dobbs, C. Roxburgh et al., Big data: the next frontier for innovation, competition, and productivity (2011)

29. R.E. Bryant, Data-intensive scalable computing for scientific applications. Comput. Sci. Eng. **13**, 25–33 (2011)

30. B. Ma, A. Shoshani, A. Sim, K. Wu, Y. Byun, J. Hahm et al., Efficient attribute-based data access in astronomy analysis, in *High Performance Computing, Networking, Storage and Analysis (SCC), 2012 SC Companion* (2012), pp. 562–571

31. V. Semkova, N. Otuka, S.P. Simakov, V. Zerkin, Experimental nuclear reaction data collection EXFOR, in *2011 2nd International Conference on Advancements in Nuclear Instrumentation Measurement Methods and their Applications (ANIMMA)* (2011), pp. 1–3

32. F.E. Lindsay, J.R.G. Townshend, J. Jaja, J. Humphries, C. Plaisant, B. Shneiderman, Developing the next generation of earth science data systems: the global land cover facility, in *Geoscience and Remote Sensing Symposium, 1999. IGARSS'99 Proceedings. IEEE 1999 International*, vol. 1 (1999), pp. 616–618

33. J. Gantz, D. Reinsel, The digital universe decade-are you ready. IDC White Paper (2010)

34. D. Laney, 3D data management: controlling data volume, velocity and variety. META Group Research Note, vol. 6 (2001)

35. W. Tian, Y. Zhao, 4—cloud resource scheduling strategies, in *Optimized Cloud Resource Management and Scheduling*, ed. by W. Tian, Y. Zhao (Morgan Kaufmann, Boston, 2015), pp. 79–93

36. T. Vidick, Three-player entangled XOR games are NP-hard to approximate, in *2013 IEEE 54th Annual Symposium on Foundations of Computer Science (FOCS)* (2013), pp. 766–775

37. K. Lakshmanan, S. Kato, R. Rajkumar, Scheduling parallel real-time tasks on multi-core processors, in *Real-Time Systems Symposium (RTSS), 2010 IEEE 31st* (2010), pp. 259–268

38. J. Dean, S. Ghemawat, MapReduce: simplified data processing on large clusters. Commun. ACM **51**, 107–113 (2008)

39. J. Gantz, D. Reinsel, Extracting value from chaos. IDC iview 9–10 (2011)

40. M. Cooper, P. Mell, Tackling big data, in *Federal Computer Security Managers' Forum* (2012)

41. S.D. Kahn, On the future of genomic data. Science **331**, 728–729 (2011)

42. K. Begeman, A. Belikov, D. Boxhoorn, F. Dijkstra, H. Holties, Z. Meyer-Zhao et al., LOFAR information system. Futur. Gener. Comp. Syst. **27**, 319–328 (2011)

43. Z. Mark, Astronomy data bounty spurs debate over access. Nature **514**, 18 (2014)

44. P. Quinn, Big data for big astronomy. Australas. Sci. **36**, 37–39 (2015)

45. E. Strohmaier, H.W. Meuer, J. Dongarra, H.D. Simon, The TOP500 list and progress in high-performance computing. Computer **48**, 42–49 (2015)

46. L.D. Stein, The case for cloud computing in genome informatics. Genome Biol. **11**, 207 (2010)

47. R. Banalagay, K.J. Covington, D.M. Wilkes, B.A. Landman, Resource estimation in high performance medical image computing. Neuroinformatics **12**, 563–573 (2014)

48. M.L. da Silva, J. Roca-Piera, J.-J. Fernández, High performance computing approaches for 3D reconstruction of complex biological specimens. Adv. Exp. Med. Biol. **680**, 335–342 (2010)

49. N. Sadashiv, S.D. Kumar, Cluster, grid and cloud computing: a detailed comparison, in *2011 6th International Conference on Computer Science & Education (ICCSE)* (2011)

50. N. Bessis, E. Asimakopoulou, T. French, P. Norrington, F. Xhafa, The big picture, from grids and clouds to crowds: a data collective computational intelligence case proposal for managing disasters, in *2010 International Conference on P2P, Parallel, Grid, Cloud and Internet Computing (3PGCIC)* (2010)

51. T. Bicer, D. Chiu, G. Agrawal, Time and cost sensitive data-intensive computing on hybrid clouds, in *2012 12th IEEE/ACM International Symposium on Cluster, Cloud and Grid Computing (CCGrid)* (2012)

52. S. Chaisiri, L. Bu-Sung, D. Niyato, Profit maximization model for cloud provider based on Windows Azure platform, in *2012 9th International Conference on Electrical Engineering/Electronics, Computer, Telecommunications and Information Technology (ECTI-CON)* (2012)

53. D. Chappell, Windows HPC server and windows azure high-performance computing in the cloud (2011)

54. T. Chen, S. Zhou, Classify virtualization strategy in cloud computing, in *2012 7th International Conference on Computer Science & Education (ICCSE)* (2012)

55. J.A. Delmerico, N.A. Byrnes, A.E. Bruno, M.D. Jones, S.M. Gallo, V. Chaudhary, Comparing the performance of clusters, Hadoop, and Active Disks on microarray correlation computations, in *2009 International Conference on High Performance Computing (HiPC)* (2009)

56. J. Ekanayake, T. Gunarathne, J. Qiu, Cloud technologies for bioinformatics applications. IEEE Trans. Parallel Distrib. Syst. **22**(6), 998–1011 (2011). https://doi.org/10.1109/tpds.2010.178

57. T. Gunarathne, Z. Bingjing, W. Tak-Lon, J. Qiu, Portable parallel programming on cloud and HPC: scientific applications of Twister4Azure, in *2011 Fourth IEEE International Conference on Utility and Cloud Computing (UCC)* (2011)

58. A. Gupta, D. Milojicic, Evaluation of HPC applications on cloud, in *Open Cirrus Summit (OCS)* (2011)

59. P. Heinzlreiter, M.T. Krieger, I. Leitner, Hadoop-based genome comparisons, in *2012 Second International Conference on Cloud and Green Computing (CGC)* (2012)

60. R. Longbottom, *Linux PC Benchmarks* [Online] (2012). Available: http://www.roylongbottom.org.uk/linux%20benchmarks.htm

61. J. Ahrens, Increasing scientific data insights about exascale class simulations under power and storage constraints. IEEE Comput. Graph. Appl. **35**, 8–11 (2015)

62. E. Afgan, B. Chapman, M. Jadan, V. Franke, J. Taylor, Using cloud computing infrastructure with CloudBioLinux, CloudMan, and Galaxy. Curr. Protoc. Bioinf. Chapter 11, Unit11.9 (2012)

63. S.V. Angiuoli, M. Matalka, A. Gussman, K. Galens, M. Vangala, D.R. Riley et al., CloVR: a virtual machine for automated and portable sequence analysis from the desktop using cloud computing. BMC Bioinf. 12, 356 (2011)

64. S.V. Angiuoli, J.R. White, M. Matalka, O. White, W.F. Fricke, Resources and costs for microbial sequence analysis evaluated using virtual machines and cloud computing. PLoS ONE 6, e26624 (2011)

65. B.G. Batista, J.C. Estrella, C.H.G. Ferreira, D.M.L. Filho, L.H.V. Nakamura, S. Reiff-Marganiec et al., Performance evaluation of resource management in cloud computing environments. PLOS ONE 10, e0141914 (2015)

66. G.B. Berriman, E. Deelman, G. Juve, M. Rynge, J.-S. Vöckler, The application of cloud computing to scientific workflows: a study of cost and performance. Philos. Trans. R. Soc. A: Math. Phys. Eng. Sci. 371, 20120066 (2012)

67. I. Bildosola, R. Río-Belver, E. Cilleruelo, G. Garechana, Design and implementation of a cloud computing adoption decision tool: generating a cloud road. PLOS ONE 10, e0134563 (2015)

68. H. Bolouri, R. Dulepet, M. Angerman, Menu-driven cloud computing and resource sharing for R and bioconductor. Bioinformatics 27, 2309–2310 (2011)

69. N. Bressan, A. James, C. McGregor, Integration of drug dosing data with physiological data streams using a cloud computing paradigm, in 2013 35th Annual International Conference of the IEEE Engineering in Medicine and Biology Society (EMBC), vol. 2013 (2013), pp. 4175–4178

70. N. Drake, Cloud computing beckons scientists. Nature 509, 543–544 (2014)

71. J.-P. Ebejer, S. Fulle, G.M. Morris, P.W. Finn, The emerging role of cloud computing in molecular modelling. J. Mol. Graph. Modell. 44, 177–187 (2013)

72. G.C. Kagadis, C. Kloukinas, K. Moore, J. Philbin, P. Papadimitroulas, C. Alexakos et al., Cloud computing in medical imaging. Med. Phys. 40, 070901 (2013)

73. I. Kim, J.-Y. Jung, T.F. Deluca, T.H. Nelson, D.P. Wall, Cloud computing for comparative genomics with windows azure platform. Evol. Bioinf. 8, 527–534 (2012)

74. M. Lawrenz, D. Shukla, V.S. Pande, Cloud computing approaches for prediction of ligand binding poses and pathways. Sci. Rep. 5, 7918 (2015)

75. H. Miras, R. Jiménez, C. Miras, C. Gomà, CloudMC: a cloud computing application for Monte Carlo simulation. Phys. Med. Biol. 58, N125-33 (2013)

76. J. Qiu, J. Ekanayake, T. Gunarathne, J.Y. Choi, S.-H. Bae, H. Li et al., Hybrid cloud and cluster computing paradigms for life science applications. BMC Bioinf. 11(Suppl 12), S3 (2010)

77. M.C. Schatz, B. Langmead, S.L. Salzberg, Cloud computing and the DNA data race. Nat. Biotechnol. 28, 691–693 (2010)

78. T.-Y. Tsai, K.-W. Chang, C.Y.-C. Chen, iScreen: world's first cloud-computing web server for virtual screening and de novo drug design based on TCM database@Taiwan. J. Comput. Aided Mol. Design 25, 525–531 (2011)

79. S. Yazar, G.E.C. Gooden, D.A. Mackey, A.W. Hewitt, Benchmarking undedicated cloud computing providers for analysis of genomic datasets. PLoS ONE 9, e108490 (2014)

80. S. Zhao, K. Prenger, L. Smith, T. Messina, H. Fan, E. Jaeger et al., Rainbow: a tool for large-scale whole-genome sequencing data analysis using cloud computing. BMC Genomics 14, 425 (2013)

81. J. Hutchinson, Curtin Trials DNA Sequencing in Azure Retrieved [Online] (2011). Available: http://www.itnews.com.au/News/271076,curtin-trials-dna-sequencing-in-azure.aspx

82. S. Kailasam, N. Gnanasambandam, J. Dharanipragada, N. Sharma, Optimizing ordered throughput using autonomic cloud bursting schedulers. Softw. Eng. IEEE Trans. 39(11), 1564–1581 (2013). https://doi.org/10.1109/tse.2013.26

83. E. Kijsipongse, U. Suriya, Scaling HPC clusters with volunteer computing for data intensive applications, in *2013 10th International Joint Conference on Computer Science and Software Engineering (JCSSE)* (2013)

84. A. Leivadeas, C. Papagianni, S. Papavassiliou, Efficient resource mapping framework over networked clouds via iterated local search-based request partitioning. IEEE Trans. Parallel Distrib. Syst. **24**(6), 1077–1086 (2012). https://doi.org/10.1109/tpds.2012.204

85. P. Makris, D.N. Skoutas, P. Rizomiliotis, C. Skianis, A user-oriented, customizable infrastructure sharing approach for hybrid cloud computing environments, in *2011 IEEE Third International Conference on Cloud Computing Technology and Science (CloudCom)* (2011)

86. G. Mateescu, W. Gentzsch, C.J. Ribbens, Hybrid computing—where HPC meets grid and cloud computing. Futur. Gener. Comput. Syst. **27**(5), 440–453. http://dx.doi.org/10.1016/j.future.2010.11.003

87. B.R. Nanjesh, K.S.V. Kumar, C.K. Madhu, G.H. Kumar, MPI based cluster computing for performance evaluation of parallel applications, in *2013 IEEE Conference on Information & Communication Technologies (ICT)* (2013)

88. R. Neumann, S. Taggeselle, R. Dumke, A. Schmietendorf, F. Muhss, A. Fiegler, Combining query performance with data integrity in the cloud: a hybrid cloud storage framework to enhance data access on the Windows Azure platform, in *2012 IEEE 5th International Conference on Cloud Computing (CLOUD)* (2012)

89. N. Nurain, H. Sarwar, M.P. Sajjad, M. Mostakim, An in-depth study of map reduce in cloud environment, in *2012 International Conference on Advanced Computer Science Applications and Technologies (ACSAT)* (2012)

90. P. Mell, T. Grance, *The NIST Definition of Cloud Computing* (Special Publication 800-145, 2011)

91. D.C. Marinescu, *Cloud Computing: Theory and Practice* (Elsevier Science, Burlington, 2013)

92. I. Bojanova, J. Zhang, J. Voas, Cloud computing. IT Prof. **15**, 12–14 (2013)

93. Nature America, Gathering clouds and a sequencing storm: why cloud computing could broaden community access to next-generation sequencing. Nature Biotechnol. **28**, 1 (2010)

94. A.M. Sabri, S.R. Balakrishnan, S.V. Moolye, C.Y. Cho, S.K. Dhillon, A.S. Sidhu, Benchmarking large scale cloud computing in Asia Pacific, in *19th IEEE International Conference on Parallel and Distributed Systems (ICPADS 2013)*, Seoul, Korea, 2013, pp. 693–698

95. A.S. Sidhu, S.R. Balakrishnan, S.K. Dhillon, HPC+Azure environment for bioinformatics applications, in *2013 IEEE International Conference on Bioinformatics and Biomedicine (BIBM)* (2013), pp. 12–15

96. Q.H. Vu, R. Asal, Legacy application migration to the cloud: practicability and methodology, in *2012 IEEE Eighth World Congress on Services (SERVICES)* (2012)

97. A. Rajan, B.K. Joshi, A. Rawat, R. Jha, K. Bhachavat, Analysis of process distribution in HPC cluster using HPL, in *2012 2nd IEEE International Conference on Parallel Distributed and Grid Computing (PDGC)* (2012)

98. Y. Simmhan, C. Van Ingen, G. Subramanian, J. Li, Bridging the gap between desktop and the cloud for escience applications, in *2010 IEEE 3rd International Conference on Cloud Computing (CLOUD)* (2010)

99. J. Simons, HPC cloud bad; HPC in the cloud good, in *2013 IEEE 27th International Symposium on Parallel & Distributed Processing (IPDPS)* (2013)

100. D. Talia, Clouds for scalable big data analytics. Computer **46**(5), 98–101 (2013). https://doi.org/10.1109/mc.2013.162

101. W.T. Tsai, G. Qi, Y. Chen, A cost-effective intelligent configuration model in cloud computing, in *2012 32nd International Conference on Distributed Computing Systems Workshops (ICDCSW)* (2012)

102. W. Zeng, J. Zhao, M. Liu, Several public commercial clouds and open source cloud computing software, in *2012 7th International Conference on Computer Science & Education (ICCSE)* (2012)

103. M. Wilken, A.W. Colombo, Benchmarking cluster- and cloud-computing deciding to outsource or not data processing in industrial applications, in *2012 IEEE International Symposium on Industrial Electronics (ISIE)* (2012)

104. W.C. Chung, C.J. Hsu, K.C. Lai, K.C. Li, Y.C. Chung, Direction-aware resource discovery service in large-scale grid and cloud computing, in *2011 IEEE International Conference on Service-Oriented Computing and Applications (SOCA)* (2011)

105. T. Hey, J. Papay, G. Pápay, *The Computing Universe: A Journey Through a Revolution* (Cambridge University Press, Cambridge, 2014)

106. M. AbdelBaky, M. Parashar, K. Hyunjoo, K.E. Jordan, V. Sachdeva, J. Sexton et al., Enabling high-performance computing as a service. Computer **45**, 72–80 (2012)

107. P. Makris, D.N. Skoutas, P. Rizomiliotis, C. Skianis, *A User-Oriented, Customizable Infrastructure Sharing Approach for Hybrid Cloud Computing Environments* (2011), pp. 432–439

108. G. Hamilton, D. Pezaros, *A Service-Oriented Measurement Infrastructure for Cloud Computing Environments* (2012), pp. 947–952

109. Y.C. Lee, C. Wang, A.Y. Zomaya, B.B. Zhou, Profit-driven scheduling for cloud services with data access awareness. J. Parallel Distrib. Comput. **72**(4), 591–602 (2012)

110. J. Li, M. Qiu, Z. Ming, G. Quan, X. Qin, Z. Gu, Online optimization for scheduling preemptable tasks on IaaS cloud systems. J. Parallel Distrib. Comput. **72**(5), 666–677 (2012)

111. S. Su, J. Li, Q. Huang, X. Huang, K. Shuang, J. Wang, Cost-efficient task scheduling for executing large programs in the cloud. Parallel Comput. **39**(4), 177–188 (2013)

112. S.S. Manvi, G.K. Shyam, Resource management for infrastructure as a service (IaaS) in cloud computing: a survey. J. Netw. Comput. Appl. **41**(5), 424–440 (2014)

113. F. Teng, F. Magoulès, A new game theoretical resource allocation algorithm for cloud computing, in *Advances in Grid and Pervasive Computing*, vol. 6104, ed. by P. Bellavista, R.-S. Chang, H.-C. Chao, S.-F. Lin, P.A. Sloot (Springer, Berlin, 2010), pp. 321–330

114. B. Sotomayor, R.S. Montero, I.M. Llorente, I. Foster, Virtual infrastructure management in private and hybrid clouds. Internet Comput. IEEE **13**, 14–22 (2009)

115. Z. Qi, Z. Quanyan, R. Boutaba, Dynamic resource allocation for spot markets in cloud computing environments, in *2011 Fourth IEEE International Conference on Utility and Cloud Computing (UCC)* (2011), pp. 178–185

116. R. Bhowmik, A. Kochut, K. Beaty, Managing responsiveness of virtual desktops using passive monitoring, in *2009. IM'09. IFIP/IEEE International Symposium on Integrated Network Management* (2009), pp. 319–326

117. Q. Zhu, G. Agrawal, Resource provisioning with budget constraints for adaptive applications in cloud environments, in *Presented at the Proceedings of the 19th ACM International Symposium on High Performance Distributed Computing*, Chicago, Illinois, 2010

118. D. Ta Nguyen Binh, L. Xiaorong, R.S.M. Goh, A framework for dynamic resource provisioning and adaptation in IaaS clouds, in *2011 IEEE Third International Conference on Cloud Computing Technology and Science (CloudCom)* (2011), pp. 312–319

119. H. Yun, M. Zafer, L. Kang-Won, D. Verma, K.K. Leung, On the mapping between logical and physical topologies, in *Communication Systems and Networks and Workshops, 2009. COMSNETS 2009. First International* (2009), pp. 1–10

120. C. Yang, W. Tianyu, L. Jianxin, An efficient resource management system for on-line virtual cluster provision, in *CLOUD'09. IEEE International Conference on Cloud Computing, 2009* (2009), pp. 72–79

121. U.S. Department of Energy, *Scientific Grand Challenges: Architectures and Technology for Extreme Scale Computing* (Dec 2009)

122. U.S. Department of Energy, *Exascale and Beyond: Configuring, Reasoning, Scaling* (Aug 2011)

123. P. Kogge, K. Bergman, S. Borkar, D. Campbell, W. Carlson, W. Dally et al., *Exascale Computing Study: Technology Challenges in Achieving Exascale Systems* (28 Sept 2008)

124. D.E. Keyes, L.C. McInnes, C. Woodward, W. Gropp, E. Myra, M. Pernice et al., Multiphysics simulations. Int. J. High Perform. Comput. Appl. **27**, 4–83 (2013)

125. J. Dongarra, J. Hittinger, J. Bell, L. Chacon, R. Falgout, M. Heroux et al., *Applied Mathematics Research For Exascale Computing* (Lawrence Livermore National Laboratory (LLNL), Livermore, CA, 2014)

126. OpenStack, *OpenStack Installation Guide for Ubuntu 14.04* (2013). Available: http://docs. openstack.org/juno/install-guide/install/apt/content/

127. D. Dudkowski, B. Tauhid, G. Nunzi, M. Brunner, *A Prototype for In-Network Management in NaaS-Enabled Networks* (2011), pp. 81–88

128. J.V.D. Berg, *Windows Azure Pack for Windows Server 2012 R2 Guide* (2015). Available: https://gallery.technet.microsoft.com/Windows-Azure-Pack-for-3332c6b9/file/124838/2/ Windows%20Azure%20Pack%20for%20Windows%20Server%202012%20R2%20Guide %20V1.pdf

129. R. Calheiros, R. Ranjan, A. Beloglazov, R. Buyya, CloudSim: a toolkit for modeling and simulation of cloud computing environments and evaluation of resource provisioning algorithms. Softw. Pract. Experience **41**, 23–50 (2011)

130. S. Boisvert, F. Laviolette, J. Corbeil, Ray: simultaneous assembly of reads from a mix of high-throughput sequencing technologies. J. Comput. Biol. **17**, 1519–1533 (2010)

131. R. Li, Y. Li, K. Kristiansen, J. Wang, SOAP: short oligonucleotide alignment program. Bioinformatics **24**, 713–714 (2008)

132. R. Luo, B. Liu, Y. Xie, Z. Li, W. Huang, J. Yuan et al., SOAPdenovo2: an empirically improved memory-efficient short-read de novo assembler. GigaScience **1**, 1–6 (2012)

133. B.G. Institute, *Short Oligonucleotide* (Beijing Genomics Institute, 2007) [Online]. Available: http://soap.genomics.org.cn/#pub2

134. R. Luo, T. Wong, J. Zhu, C.-M. Liu, X. Zhu, E. Wu et al., SOAP3-dp: fast, accurate and sensitive GPU-based short read aligner. PLoS ONE **8**, e65632 (2013)

Printed in the United States
By Bookmasters